Good Tidings

Good Tidings

The History and Ecology of Shellfish Farming in the Northeast

Barbara Brennessel

UNIVERSITY PRESS OF NEW ENGLAND

Hanover and London

Published by University Press of New England,
One Court Street, Lebanon, NH 03766
www.upne.com

© 2008 by University Press of New England

Printed in the United States of America
5 4 3 2 1

Library of Congress Cataloging-in-Publication Data
Brennessel, Barbara.
Good tidings : the history and ecology of shellfish farming
in the Northeast / Barbara Brennessel.
p. cm.
Includes bibliographical references and index.
ISBN 978-1-58465-727-9 (cloth : alk. paper)
1. Shellfish culture—New England. I. Title
SH365.N35B74   2008
639.40974—dc22   2008040274

University Press of New England is a member of the Green Press Initiative. The paper used in this book meets their minimum requirement for recycled paper.

CONTENTS

PREFACE

When my husband and I moved to New England in 1977 and began to spend time in Wellfleet, on Cape Cod, one of our first purchases was a clam rake. During summers on Long Island, we had always dug clams with our feet, but we soon found out that this primitive harvesting method did not transfer well to the Cape. Sharp oyster shells on tidal flats and just beneath the sediment made it very hazardous to toe for clams and made the clam rake an absolute necessity. With noncommercial permits attached to our hats, we would spend a delightful hour, scratching the flats for quahogs and leaving muddy holes as prints on the tide.

As the years went by, it became more and more difficult to find clams in areas where they were once abundant. It was the same old story: too many people and not enough clams. Since that time, the town's Shellfish Department began a cultivation program to support a "put and take" fishery for those of us who are recreational clammers; the department grows the clams and scatters them around, and we dig them up and take them home.

The situation for the commercial shellfishing industry on the Cape paralleled the experience of recreational clammers. Too many shellfishermen were harvesting too many shellfish; clams and oysters were declining rapidly. Several recommendations were implemented by local shellfish departments, including the closure of certain shellfish beds for specific months of the year. There was renewed interest in obtaining a special license for the planting and propagation of shellfish in specific locations. More and more, wild shellfishermen become shellfish farmers.

Many individuals rely heavily on these animals for their livelihood. The importance of shellfishing on the lower Cape is evident from the day-to-day posting of reports and information about the status of the fishery in regional newspapers and on town websites. Every event, natural or otherwise, that affects the industry is big news: red tides, storms, water quality, erosion, disputes over tidal flats, tidal restrictions, and so forth.

As a biologist who studies the diamondback terrapin, a brackish water turtle under state protection in Massachusetts, I was also drawn into the shellfishing and aquaculture realm. To conduct my research on terrapins, I needed access to certain creeks that were bordered by licensed shellfishing areas. One of the first times I tried to make my way to my study area, bucket and net in hand, I was accosted by a shellfish farmer and told, in no uncertain terms, to get off his property. After my initial shock, I realized that my bucket led him to think that I was setting out to pick up some of his oysters. I hastily explained my mission and he gave me an overview of his operation and showed me how I could get to the water without interfering with his business. I thought he was doing me a favor, but later I found out that I have every right to pass through a licensed shellfish area unless it is privately owned. Over the course of several field seasons, my relationship with the shellfish farmer became more cordial, to the point that he would always give me a wave and even offer me rides in his pickup so I wouldn't have to trudge out over the flats with all my equipment. This experience made it apparent that I had a lot to learn about an industry that was taking place in a habitat shared not only with a threatened reptile, but with other folks who wanted to use the tidelands. Growing shellfish in licensed areas has saved the commercial shellfishing industry, but has not been without its problems.

Recently, several of my neighbors and friends have gotten into the shellfish aquaculture business. From the perspective of a biologist, I became fascinated with their operations and wanted to know more about the animals that were being raised and their interactions with the environment. I wondered about the effect of this industry on other creatures that shared space with the shellfish farms and on the tidal flats in general, and had a suspicion that other industry outsiders were wondering the same thing. I began to attend shellfish conferences, meetings, courses, and workshops, and to talk to those who were making a living on the flats. I collected my findings, impressions, some historical information, and a few recipes, and I share them in this book with others who may be curious about the flurry of activity that occurs on the flats at low tide.

ACKNOWLEDGMENTS

In compiling the information and resources used for this book, I consulted a multitude of experts. For their insights, knowledge, and patience with my innumerable questions, I thank Dr. Roxanna Smolowitz, whose laboratory was at the Marine Biological Laboratory in Woods Hole when I interviewed her; Dr. Dale Leavitt from Roger Williams University; Dr. William "Bill" Walton from the Cape Cod Cooperative Extension; and Dr. Kim Tetrault from the Cornell Cooperative Extension in Southold, New York. I am indebted to Dave Alves of Rhode Island, Tessa Getchis of Connecticut, and Samantha Horn-Olsen from Maine, who educated me about rules and regulations in their respective states. Cape Codders who answered questions and provided leads and insights include Bethany Walton, Diane Murphy, Sandy Macfarlane, and members of the Wellfleet Shellfish Department, Andy Koch and John Mankevetch. Assistant Constable Mankevetch allowed me to accompany him on his daily rounds and gave me a tour of the town shellfish nursery and grow-out beds.

Helen Miranda Wilson of Wellfleet supplied copies of all the historic reports of the Wellfleet Shellfish Department. Many aquacuturists allowed me to visit their "grants," take photos, ask questions, and gave me first-hand accounts of their culture techniques. In particular, I wish to thank Russ Junkins, Paul Bonanno, Barbara Austin, Jim O'Connell, Nate Johnson, and John Lovell, who all were a friendly presence on the tide, always willing to provide perspectives on the current issues and solutions to the problem of the day.

Tom Gerhardt, caretaker and interpreter at Aptucxet Trading Post in Bourne, provided resources about wampum; Judy Wilson, from the Snow Library in Orleans, Massachusetts, helped to obtain a copy of a print in the library collection; and Hope Morrill, archivist at the Cape Cod National Seashore, provided me with a wealth of information from the digitized materials in the National Park collection. Brian Bowes of Coastal Aquacultural Supply provided a wonderful overview of aquaculture practices and techniques when I toured his warehouse in Cranston.

Others who granted interviews or allowed me to probe their perspectives about the shellfish aquaculture industry were Dr. Tim Scott of Roger

Williams University and John Baldwin, Seafood Divers, Inc. Captain Andrew Cummings of South Wellfleet, whose photo appears on the back cover, and Dr. Scott Shumway of Wheaton College generously provided photos. My daughters, Nina and Adriana Picariello, took photos when I needed them, and Marisa Picariello picked up pen and ink when I requested a drawing or diagram. My husband, Nick Picariello, has scratched the flats with me for over forty years and accompanied me on several of my fact-finding trips and site visits when he could take a break from his medical practice. John Fadden supplied digitized photos of his wampum belt artwork, and aquaculturist, Heidi Gallo from Eastham, Massachusetts, provided the watercolor that appears on the front cover.

Kathy Rogers from Wheaton College was indispensable and I should have thanked her on a daily basis for continuous help with the manuscript and figures. I am greatly indebted to Sandy Macfarlane, an experienced and well-respected shellfish expert, for her careful reading of this manuscript and for her constructive comments and valuable suggestions. The book would not have materialized without the encouragement, guidance, and efforts of my editor, Richard Pult and the phenomenal assistance of the staff at University Press of New England.

In some ways, the writing of this book was similar to focusing on a moving target. Marine aquaculture is a rapidly advancing and constantly changing industry, and as a result, methods and regulations are constantly shifting and evolving while industry participants come and go. I have attempted to provide a look back at the origins of bivalve shellfish aquaculture, a snapshot of present conditions, and a view toward the future. I take responsibility for any omissions, misinterpretations, or inaccuracies presented in this work.

Chapter 1

Snug in Their Shells

Between late November and early December of each year, the shellfish growers of Wellfleet, Massachusetts, must make a serious decision. Their oysters, especially the ones below legal market size, must survive the winter so that they can be harvested during the next growing season. The oysters are in bags and trays, suspended on metal racks and PVC pipe, 18 inches from the bottom on licensed growing areas, commonly referred to as "grants" (fig. 1.1). If the oncoming winter is cold enough and ice forms in the harbor, the entire oyster crop could be destroyed. Cultured clams are less likely to experience problems during the cold months of the year. Under the sediment, clams will enter winter dormancy and remain protected from the large ice sheets that form and shift within the harbor. However, oysters, and the gear that is used to grow them, are in harm's way. Big tides will break the ice into chunks the size of a car and the resulting icebergs will move back and forth with each ebb and flow, scraping along the tidal flats with the potential to destroy a grower's investment. Most areas licensed for shellfish aquaculture in Wellfleet are located in vulnerable intertidal locations. The shellfish growers have few deep-water areas where their oysters will survive under the ice. Some growers watch the long-term weather forecast and remain prepared to retrieve all their gear and oysters from their "grants." By leaving their crop in the water, they have access to oysters that they can sell and are delaying the backbreaking work that is involved in removing equipment and oysters from their farms. Others are even more wary. They know how quickly weather conditions can change in New England and would rather be safe than sorry. They begin preparations to store their oysters so that the animals can sleep safely through the cold months of the year.

The location of a shellfish growing area will dictate the strategies that are available for overwintering of oysters. In deep-water areas of many northeastern locations, oysters can be placed in cages and kept on the bottom, well below the ice, a method known as wet storage. These cages are not an option in areas where dredging or dragging for oysters occurs over the win-

FIG. 1.1 Licensed shellfish aquaculture site on the tidal flats in Eastham, Massachusetts.
Photo: Barbara Brennessel

ter or in areas near navigation channels where the oysters could be disturbed and stressed. Access to oysters is also important for farmers who rely on a year-round supply to satisfy their customers. If the oysters are in deep water, a grower might put them in a place where they can be harvested from time to time. A shellfish grower from Duxbury, Massachusetts, has been known to use his chain saw to retrieve market oysters from their icy domain.

If deep-water sites are not available, the oyster farmer can opt for dry storage under cold conditions. Unlike other bivalves, oysters can survive out of the water, not only between tides, but for months at a time, as long as the air is cold, humid, and above freezing. In the 1860s, Henry David Thoreau was amazed by the fact that Wellfleet oystermen kept their foot-long oysters—which had to be cut in two before they could be swallowed—in cellars throughout the winter. He questioned one of the local oystermen about it.

"Without anything to eat or drink?" I asked.
"Without anything to eat or drink," he answered.
"Can the oysters move?"
"Just as much as my shoe." (Thoreau [1865] 2004, 65)

Optimal temperatures for winter dry storage are between 0 and 3°C (32 and 38°F). The oyster will have the best chance of survival if it sleeps in its cup, or deeper shell; this orientation will keep the liquid in the shell and prevent desiccation. A large, cooled room or refrigerator can be used, but if a proper piece of cooling equipment is not available or if a grower has more oysters than will fit into a refrigerator, the oysters can be pitted, or put in a hole in the ground. I have been told that new, unused, cement septic tanks, buried in the ground, make excellent pitting places.

Pitting is an art in which conditions have to be optimized to prevent mortality. As a general rule, the storage pit cannot collect water nor be above the frost line. Oysters usually are bagged in burlap before pitting and the burlap is kept moist. Some oyster growers use ventilated trays that they can stack atop one another. No matter how controlled the environment of a dry-storage area, some oysters will die, and mortality will increase the longer the oysters remain in storage. It will be early springtime, when the risk of ice has passed, before many Wellfleet growers will know how much of their oyster crop has survived through the winter.

Origins of Bivalve Shellfish Aquaculture

The major impetus for the movement of the shellfish industry from wild harvest to aquaculture was the depletion of wild shellfish from their historic natural locations. The demand for shellfish, coupled with overharvest of the resource, shellfish diseases, pollution, and degradation of shellfish habitat, have led to global declines of many economically important bivalve mollusc species. Although wild harvest is still important to shellfisheries in some areas, it is not surprising to find that maintaining a viable shellfish industry also involves attempts to grow shellfish under somewhat artificial or managed conditions. Growing oysters under circumstances that require pitting or other forms of winter storage may seem strange to those of us who have never given much thought to the work involved in bringing those oysters to the raw bar or restaurant. In reality, most of the oysters that are consumed on the half shell are available to us as a result of the work of oyster growers. The labor-intensive method of growing oysters and other shellfish is known as shellfish aquaculture, shellfish farming, or shellfish mariculture. This endeavor requires a considerable amount of human intervention, including transplantation, stocking, seeding, breeding, feeding, protecting, and an extensive level of oversight by regulatory agencies. The culture of bivalve shellfish has been practiced in some form or another since antiquity and in the United States since soon after the arrival of European colonists.

Whether harvesting pearl or edible oysters in East Asia, mussels, oysters, and cockles in Europe, mussels in Canada, or oysters and clams in the United States, the success of the industry relies on some form of manipulation of shellfish resources. The movement to culture shellfish had its origins in some trial-and-error shellfish transplant efforts, some innovation to collect spat (tiny oysters, sometimes referred to as seed), and some industrious shellfishermen who experimented with new techniques.

Ancient forms of shellfish farming stemmed from observations that small oysters could become attached to broken pottery, remnants of containers that held olive oil and wine, which littered the Mediterranean and Aegean sea bottoms. Shipwrecks were responsible for providing these fortuitous spat collectors. Some enterprising fishermen transplanted the oysters, pot fragments and all, to other areas where they seemed to grow better. A Roman citizen, Sergius Orata, took the observation to new heights and made his fortune in the second century B.C. by transferring oysters from Brindisi, at Italy's heel on the Adriatic coast, to oyster ponds that he created across the country in the Gulf of Baia, near Naples (Clark 1959; Hedeen 1986). Orata's initial cultivating efforts involved a simple transfer of oysters from one area to another. Later, he devised an apparatus that could be used to rear oysters on a soft bottom and also to collect spat. A pile of rocks formed a solid substrate, while bundles of twigs attached to the rocks provided a suitable foundation for the attachment of spat (Brooks [1891] 1996). By using public lands to cultivate his transplants, Orata set up the first conflict between shellfish farmers and other citizens who used the same lakes and ponds for bathing. Throughout ancient Roman times, oyster ponds continued to be used for some forms of oyster culture; a few samples of ancient artwork depict various stick structures that probably served as spat collectors (Clark 1959).

In several parts of the world, techniques were developed to raise shellfish from spat. The Chinese and Japanese used stones, bamboo, and various floating devices to collect oyster spat. In Europe, tiny mussels were collected on nets secured to the bottom with wooden pickets. In their attempts at oyster cultivation, Europeans tried various devices to collect spat. These contraptions initially were made from ropes and sticks, but eventually tiles were employed. In Mexico and Central America, tree branches were placed in the water to attract passing seed oysters.

The French developed a substantial wild oyster fishery but for reasons that are unclear, the industry collapsed in the mid-nineteenth century. Historians conjecture that an unidentified disease may have been responsible.

Oysters suddenly were sparse from Normandy to Arcachon near Bordeaux, areas in which they once were abundant. In the 1850s, the French government hired Professor Coste of the College of France to save the country's oysters. In attempts to revitalize the industry, Coste was one of the first to undertake a systematic study of oyster propagation. He used half-round roofing tiles covered with a lime/sand mixture to catch oyster spat. The spat was removed from the tiles and transferred to an "ambulance," a structure that we would refer to as a stretcher: a wooden raft with handles and legs, covered with a net. The oysters were grown out on the "ambulance" to approximately 1.25 cm (0.5 inch). They would then be moved to different areas to fatten up and achieve distinctive regional flavors (Clark 1959; Hedeen 1986).

As a result of Coste's findings, the industry in Normandy adopted techniques involving the cultivation of transplanted oysters in parks. The parks consisted of collections of basins, called claires, where oysters fed on algae and grew quickly. There were no restrictions on the use of public waters, so early oyster cultivators set up shop in near-shore basins that had previous usage for salt production; they later extended their parks to deeper waters. Some of the oysters developed a greenish hue to their meat, probably from the algae they filtered, but some green coloration also may have been imparted by the contents of the sediment or by copper in the water (Hedeen 1986). Despite the color, green oysters became a much desired delicacy. Dirt walls were built up around the claires so that water was exchanged only at the highest tides. This also allowed farmers to regulate water level in the claires to compensate for excessive heat or cold. A farmer in Marennes typically could culture 150,000 oysters in a 2-acre claire (Brooks [1891] 1996).

In the 1860s, French oyster growers largely replaced the disappearing oysters that were native to western France, *Ostrea edulis* with Portugese oysters, *Crassostrea angulata.* Over the years, the industry has had its ups and downs as a result of pollution or disease. In modern French oyster farms, tiles are still used to catch oyster spat and the basic system for aquaculture that was developed by Coste is still employed. Other oyster species, such as the Pacific oyster, *Crassostrea gigas,* have been introduced and now form the bulk of the industry.

On the other side of the Atlantic, the early colonists learned from the natives the value of oysters and several varieties of clam as a food source. During the warm months of the year, when coastal tribes camped out on the beaches, the natives would wade on tidal flats to find oysters and they also would dig through the sediments with hands and feet to feel for clams. Clams and oysters were consumed in great quantities. Sometimes they were baked in great, rock-lined pits along with corn and seaweed, a precursor to the modern clambake. As the clambake evolved and was refined by settlers, soft-shelled clams replaced the hard clams or quahogs. The smaller quahogs were eaten raw, while the larger ones were used to make chowders. The natives also hung shellfish meats on lines to dry, saving them for winter storage or using them in trade with inland tribes. When metal tools became available, colonists fashioned rakes and other digging implements that were more efficient than bare feet for sifting sediments for clams.

When Samuel de Champlain sailed into Wellfleet Harbor on Cape Cod in 1606, during his second expedition, the harbor was so chock full of oysters that he named it Port aux Huitres, that is, Port of Oysters or Oyster Harbor. However, it didn't take very long for the situation to change and for the expansive beds of oysters to disappear. Overharvesting was the underlying factor that led to the rapid depletion of oysters within 150 years of the original settlement of the area by Europeans. In his history of Cape Cod, Scott Corbett writes about the bounty of shellfish and describes oysters as one of the treasures of the Cape (Corbett 1955). Their depletion by colonists necessitated legislation to protect the resource, yet the native beds were decimated by the 1770s. Corbett recognized that the final disappearance of oysters was due to overharvest, but he also posits that disease or parasites may have factored into their precipitous decline, a possibility that was never confirmed. Some Wellfleet citizens believed the disease was "sent by Providence upon the oysters, as a punishment for the sins of the fishermen, who certainly were more worthy of such an affliction than the helpless oysters" (Brooks [1891] 1996, 176).

While writing about the wholesale exploitation of shellfish on Cape Cod, Corbett also commented on the overharvest of hard clams. The local clams, which later became known as quahogs, were used for both food and bait. At the beginning of the nineteenth century, a few hundred Cape Codders were able to make their living by digging and opening clams. The going rate was three dollars a barrel and it was possible to fill a barrel and a half in one week.

To restore declining oyster stocks, the Massachusetts colonists "borrowed" oysters from points south, even as far as Chesapeake Bay. Wellfleet replaced its declining stocks of famous "Billingsgates" with foreign oysters (fig. 1.2).

[S]omeone had the happy thought of bringing young oysters from beds further south and letting them mature and get their flavor in the superior waters of Cape Cod. The experiment was a complete success. The

FIG. 1.2 Southern trade: A ship bringing oysters from Chesapeake Bay to replenish oysters stocks in the Northeast, circa late 1800s. *Photo: Courtesy of the National Park Service, Cape Cod National Seashore*

newcomers turned out to be as good as even the native oysters had been; it is to be hoped that Cape Cod can do as much with the human new-comer. Of course, with the oyster, its place of birth did not matter—it was the finishing school that counted. By 1830, the oyster was enjoying a far greater vogue than it enjoys even today, and it formed a part of the Cape's economy worthy of notice." (Corbett 1955, 55)

Some oysters were shucked by local families, but larger-scale operations were conducted in oyster houses that sprung up along Wellfleet Harbor, many of them built on stilts to allow the oyster boats to pull up and un-load (fig. 1.3). From the shacks, the oysters were packaged and shipped whole or as shucked meats. Boston became a big oyster town due to Cape Cod oyster supplies, especially those from Wellfleet, supplemented with shellfish from the Charles and Mystic rivers and nearby Portsmouth, New Hampshire. Boston's oyster houses became meeting places for politicians and gathering spots for the social elite. With the continued demand from Boston, Wellfleet remained a thriving shellfish harvest area throughout the nineteenth century, with many oyster and clam shacks lining the shore so that shellfish could be rapidly off-loaded and shipped to Boston. One of the most famous of these shacks was owned by Cap'n Higgins and came to be known as the "Spit and Chatter Club." During foul weather, the local

FIG. 1.3 Wellfleet oyster house, circa early 1900s. *Photo: Courtesy of the National Park Service, Cape Cod National Seashore*

oyster harvesters would gather to repair gear, swap stories, and play cards around the wood-burning stove.

As settlers to New England discovered, coastal areas in Rhode Island, especially in Narragansett Bay, had abundant oysters, which colonial newcomers observed the natives harvesting by hand. Aware of the importance of the resource, King Charles II provided the rights of free fishing to the Colony of Rhode Island in 1683 (Kochiss 1974). For the early Rhode Island colonists, oysters became a food item that provided sustenance while they developed land for farming. Soon, Rhode Island oysters were being taken not just for the family table by individuals. A few enterprising oystermen began to dredge for oysters and sell them to populations in towns and villages that were distant from the shore. The dredges consisted of wire baskets with teeth that were pulled along the bottom by boat and lifted with a winch (fig. 1.4). Over time, the resource was being eroded by continued, unregulated harvest. As a response to the decline in oysters, dredging by sail-powered vessels was banned in 1766, but the oysters continued to disappear, victims of another, simpler, method of harvest: tonging. A pair of tongs consists of two rakes that are fastened together like scissors (fig. 1.5). In the early years of tonging, the tonger would stand on a small dugout or canoe and could rake oysters up to depths of about fifteen feet. By 1865, a Rhode Island, shellfisherman could lease an area for $10 per

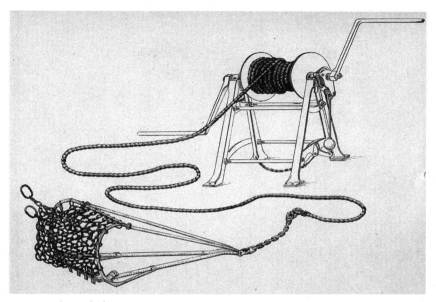

F I G . 1 . 4 Oyster dredge. *Photo: NOAA Photo Library, National Marine Fisheries Service, National Oceanic and Atmospheric Administration, U.S. Department of Commerce*

acre on any "bottoms which are covered by water at low tide and are not within any harbor line to be used as a private oyster fishery for the planting and cultivation of oysters, whether these lands contain natural beds or not." (Brooks [1891] 1996, 127–28). The state-run lease system gave ten-year rights for leaseholders. Some of the bottom areas were leased by Connecticut oyster companies under the name of Rhode Island residents (Kochiss 1974). Until 1883, Rhode Island grounds could be used only for planting seed purchased from other states, not for moving oysters from one area to another within the state.

In Connecticut, New Haven became the center of the oyster industry. The first Connecticut law to protect oysters was enacted in 1762, and by 1784, towns were allowed to regulate the harvest of oysters. In New Haven, catch limits were imposed and fines were levied against poachers. Oysters were shucked in oyster shops and in cellars of people's homes and packed in kegs with ice and sent to bars, taverns, and restaurants all over New England and New York with horse-drawn wagons. Families would bring empty kegs to meet the oyster wagons, purchase whole oysters, and bring them home to shuck them. Once the kegs were filled with oyster meats, they were sold to other buyers. Oyster boats from Cape Cod also delivered oys-

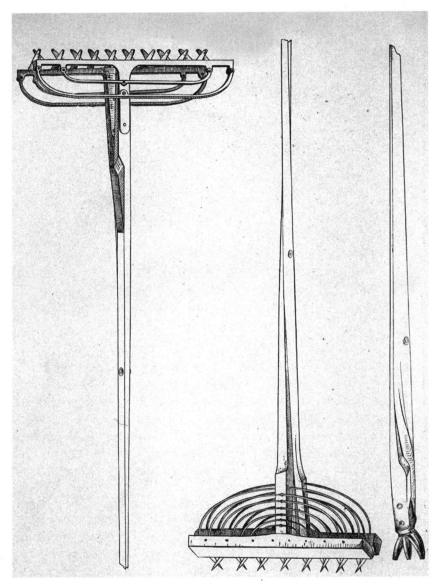

FIG. 1.5 Oyster tongs. *Photo: NOAA Photo Library, National Marine Fisheries Service, National Oceanic and Atmospheric Administration, U.S. Department of Commerce*

ters to shops, located on plants over the water where shucking operations were set up (Galpin 1989; MacKenzie 1996). Northern oysters have good keeping qualities; some were iced and shipped in the shell and usually would arrive at their destination in good condition without the loss of their juices. By 1854, when few oysters were left in state waters, Connecticut oystermen began importing oysters from New York, New Jersey, Delaware, and the Chesapeake Bay and planting them on beds marked by tree branches.

New York City was the world's oyster capital. Oysters were once abundant in the water surrounding Manhattan and a large, thriving industry revolved around this bivalve. Liberty Island was called Big Oyster Island while Ellis Island was named Little Oyster Island. Today, it is difficult to imagine the extent of the industry and the popularity of oysters. Oyster barges lined the waterfront, where they served as collection areas; the oysters were removed from boats, processed, and shipped to local and distant markets, even to Europe. New Yorkers were proud when their native oysters had name-brand recognition: Bluepoints, Rockaways, Canarsies, Jamaica Bays, City Islands, Prince's Bays, and Sounds. The name Bluepoint was so popular that the New York State legislature ruled in 1908 that no oyster could be sold as a Bluepoint unless it spent at least three months in Great South Bay (Kochiss 1974).

In addition to the natural bounty in the vicinity of Manhattan, much of the oyster harvest from northeastern states and even from Chesapeake Bay was funneled through New York. Initially, oysters arrived by schooner, but as transportation networks developed, they also were delivered by wagon and rail. Express trains made deliveries from Long Island to Manhattan seven times each day. New Yorkers loved their oysters. Not only were they consumed at family tables but they also made their way to taverns and oyster cellars throughout the city where they were served raw, on the half shell, as many as a person could eat. The clientele were known to eat a vast quantity of oysters in a sitting, not like today's appetizer of a half dozen. The oysters also were served at upscale eating establishments, such as Delmonico's in Manhattan, where they were prepared in soups and stews and also pickled and fried.

Disputes over the Bottoms

Throughout the Northeast, many conflicts arose over oyster-rich bottomlands. Early settlers, using European traditions and cultural attitudes, treated the landscape and natural resources as commodities and property that could be bought, sold, and traded.

> As early as 1679, the Long Island town of Brookhaven had passed an ordinance restricting to ten the number of vessels allowed in the Great South Bay, a huge natural oyster bed between Long Island and Fire Island. The European settlers in Brookhaven quickly realized the potential of Great South Bay Oyster beds. In 1767, the town negotiated an arrangement recognized by the King of England in which the town, in exchange for control over the oyster beds, agreed to give the crown's recognized owner, William Smith and his heirs forever half of all "net income accruing to the town from the use of the bottom of the bay." (Kurlansky 2006, 83)

In 1910, the town of Brookhaven took control of the oyster bottoms and eventually, the title of a huge tract of bottomland, 13,440 acres, was given to the Bluepoints Oyster Company, Inc., in West Sayville (Kochiss 1974).

After the War of Independence, the newly formed states treated shellfish beds differently; most states promoted free and equal rights for the public to fish and use the waters for navigation. This stance seems egalitarian, but it led to serious conflicts, especially as states deemed it necessary to regulate the industry. The notion of common use led to common problems: too many shellfishermen, too few shellfish, and the destruction of natural oyster beds. Many writers have pointed out the similarity between common use of the shellfishery—or any fishery for that matter—and the famous essay on human population growth written by Garrett Hardin about the tragedy of the commons (Hardin 1968). According to Hardin's theory, when common resources are utilized by more than one person and must be shared by increasing numbers of individuals, there is less incentive on the part of any one person to preserve the resource. The result is over-exploitation and eventual destruction or disappearance of the resource.

To protect the common right to harvest shellfish, states and towns began to regulate the industry. As early as 1719, a New Jersey act was passed that closed the shellfish beds from May 10 to September 1 of each year and restricted shellfishing in New Jersey waters to state residents (McCay 1998). Elsewhere, oyster-rich towns attempted to regulate their resource by preventing nonresidents from harvesting oysters, and by levying fines. In a landmark case brought to the New Jersey State Supreme Court in 1821, Justice Kirkpatrick essentially ruled that the state "owned" the lands under mean high water (McCay 1998). The court decision became part of current public-trust doctrine, which implies that, in New Jersey, the state holds the resource "in trust" for the public. In some northeastern states such as Massachusetts, Maine, and New York, towns retained title to shell-fishing areas unless the area was part of an original private land title. Some of these titles still exist and can be traced back to colonial times, and the rightful "owners" of a number of important intertidal properties are still disputed.

If a town or the state retains title to shellfishing areas, they act as trustees and are caretakers of the bottomlands for the public. They are entrusted with safeguarding the shellfish waters and other coastal resources for common use. However, interpretation of the nature of the protection that was afforded has varied substantially, especially as oyster stocks began to decrease. In some cases, it was not even clear which state or town owned the resource. For example, during the colonial period New York and New Jersey fought over oyster fishing rights around Staten Island, Raritan Bay, and Arthur Kill (a body of water between New Jersey and Staten Island).

Movement and Transplantation of Oysters

As oyster resources continued to be depleted, oystermen attempted to buoy the industry by devising creative methods that can be considered as precursors to modern shellfish aquaculture. Declines in oyster harvest in most Northeast locations gave rise to the strategy of transplanting oysters to formerly productive waters in the New York City area. By 1820, the beds around Staten Island, New York, were among the first to be destroyed by overharvest, prompting oystermen to move small oysters from Arthur Kill and the Raritan River in New Jersey and parts of Long Island to the depleted beds. Other local transplanting strategies were developed. In the early eighteenth century, Great South Bay, on the southern shore of Long Island, was a productive area for harvest of oysters for New York City markets. It was observed that young oysters did well in the eastern section of the bay, which was not salty enough for oyster drills, a serious predator that plagued the western part of the bay. Although drills were not present in the eastern section, the less-salty waters were not conducive to optimal oyster growth. The oyster fishermen adopted a transplanting strategy in which they grew their oysters in the eastern part of the bay until they were large enough to fend off predation by drills, then moved the oysters to the western section of the bay for further growth (Kurlansky 2006).

By 1825, the declining oyster industry in New Jersey, New York, and

New England was relying heavily on oysters that were taken by specially commissioned schooners from Chesapeake Bay and planted in northern waters. Chesapeake oysters revived the declining oyster resources in the northeast, but with the passion for oysters in full gear, Chesapeake oysters also were becoming depleted. During the nineteenth century, Chesapeake watermen from Maryland and Virginia were constantly disputing their rights to harvest oysters in various locations within the bay. The conflicts, dubbed by history as "the Oyster Wars," escalated over the rights to harvest oysters. Furthermore, Chesapeake watermen were not happy about the removal of their resources by boats dispatched by oyster companies from the north. The so-called Virginia or southern trade—transplanting of Chesapeake oysters to northern waters—was having an impact on their livelihood. The conflicts were exacerbated when dredging techniques were employed. The dredgers were much more efficient at harvesting oysters than tongs and other hand tools. But dredging has the potential to be very destructive. According to Susan Brait, "A dredge works the way a snowplow would if it were pulled instead of pushed. Clawing the ground with a set of metal teeth, it scrapes up into a net bag held open by an iron frame everything in its path. A captain drags away the rock itself, layer by layer, when he pulls a dredge across a reef again and again" (Brait 1990, 30). And Brait reminds us that "Dredging is not an absolute evil. When oysters growing undisturbed pile themselves up, the competition among them for food and space is extreme. Crowded, and probably a little hungry they become . . . snaps: long skinny oysters without much meat. Limited dredging, which thins and weeds them, gives them room to get fat" (38).

In the early 1800s, dredging was a very productive but entirely unregulated method to harvest oysters. Maryland passed the Acts of 1820 that were aimed at northeast fishermen, especially the dredgers. The acts prevented nonresidents from harvesting oysters in the Maryland section of Chesapeake Bay. This did not stop some New England oyster companies, which continued to dredge Chesapeake Bay, often during cover of night when their men would be less likely to be shot. As a result of their clandestine activities, these northern oyster pirates were known as the Mosquito Fleet.

Transplantation Provokes More Disputes over Bottomlands

The harvest of wild, naturally occurring shellfish from bottomlands was not the only contentious aspect of the growing industry. Legal issues surfaced over the movement of oysters, the transplanting of shellfish, and the use of public bottomland for commercial shellfish operations. It seems reasonable to learn that a natural oyster bed could not become a specific fisherman's property and that it is available for all who want to harvest it, providing they are willing to abide by harvest regulations. The right to harvest shellfish becomes less clear when fishermen begin to move shellfish around, taking spat or seed bivalves from one area and planting or bedding them in another area, perhaps one that contains some native shellfish. Conflicts were bound to arise between those who harvested wild stocks and those who were practicing aquaculture by planting and cultivating shellfish.

The large-scale transplanting of oysters in the Northeast raised many issues about where the oysters could be planted and who had rights to harvest the transplanted oysters. In the early days of oyster planting, any oysterman in New York could claim an underwater lot, mark it off with wooden stakes, and plant his oysters. New York and New Jersey enacted laws to protect the interests of this new breed of oyster culturists. Recognizing that they had invested time, labor, and a capital outlay into their transplanting operations, the planters were awarded exclusive harvest of the shellfish they had caught and planted. This was the beginning of the trend toward current shellfish farming practices in the Northeast. Shellfish growers were embarking on a new type of fishing operation, more akin to farming than fishing. Shellfish crops were cultivated, seed was planted, farms were tended, and the crop was harvested.

In 1854, a Connecticut resident could plant oysters from another state in Connecticut waters; by 1855, oysters could be moved within the state. A law was enacted in 1855 that allowed individual residents to lease up to 2 underwater acres for shellfish cultivation. Some business-minded individuals convinced families and friends to lease adjacent two acre plots and then consolidated them into larger tracts. In this way, Connecticut laws unintentionally allowed the development of large oyster companies by giving access to extensive planting areas to enterprising individuals. Some of the leases were expanded to deep-water areas and, in 1881, perpetual franchises could be obtained from the newly formed State Shellfish Commission. This movement essentially privatized large areas in Long Island Sound. A new category of leased area was added in 1915, when underwater areas that were deemed fallow for ten years (that is, contained no clams or oysters), could be leased from the state for ten years at a time (Kochiss 1974).

Other states and towns took different approaches to the treatment of common grounds. Rather than perpetual franchises or long-term leases, shorter leasing programs were put into place. Furthermore, to protect the rights of wild fishermen who did not have the resources to transplant oysters from other locations and thus relied on wild stocks for their livelihood, policies were adopted that allowed leases only in areas that did not already contain shellfish. Thus, areas that contained shellfish were allowed to remain part of the public shellfishing domain. As fair as this seems, it was not always easy to document which areas were barren and which contained oysters. Many disputes arose between fishermen who had planted oysters and harvesters of those planted oysters who maintained that they were working in public waters.

Some of the largest oyster companies that were established in areas where planting was permitted had their own culling (separating oysters that are clumped together and/or sorting oysters by size), shucking, and packaging operations. The R.R. Higgins Co., founded in 1828, dominated the Boston market. Other large companies that emerged were Robert Pettis of Providence, Rhode Island, in 1840, F. Mansfield and Sons Co. in New Haven in 1846, George Sill in New York in 1857, and H.C. Rowe and Co. in New Haven in the late 1850s.

Some Northeast towns took a different approach and had restrictions that were designed to protect the industry from encroachment by large corporations. The size of leased areas was limited to only a few acres, corporations were not allowed to lease bottomlands, and techniques for harvest of the planted oysters were restricted to the use of simple hand tools, "keeping shellfishing a job for people with strong arms and backs willing to work extremely hard for long hours on the water" (McCay 1998, 17). This attitude contributed to a movement to protect individuals and preserve a way of life that was valued in many fishing communities.

Oyster Culture

Transplanting was a temporary fix. Moving oysters from region to region was not a panacea for overharvesting, and was rapidly depleting the resource from historically oyster-rich areas. Removal of oysters from Chesapeake Bay by northern shellfish operations contributed to the decline in populations of Maryland and Virginia oysters. Large-scale transplantation of oysters also had the potential to move diseases and predators. To address the oyster problems in Chesapeake Bay, William K. Brooks was appointed oyster commissioner of Maryland in 1882. Brooks was a professor at Johns Hopkins University, where his research was devoted to the study of the oyster. Brooks saw the writing on the wall and was the first to propose oyster culture as a method to save the formerly booming Chesapeake Bay oyster industry from depletion of its resources. He made a study of French oyster-culture methods and reviewed the early oyster industry in the United States. His astute observations and analyses, as well as his proposals to save a dying industry, were summarized in his seminal work, *The Oyster* (Brooks [1891] 1996).

Brooks looked for additional solutions to replenish oysters in areas where they once were abundant. For hundreds of years, as oysters were harvested, so were their shells. Oyster shells were extracted for lime to use as agricultural fertilizer and to make mortar. Crushed shells were used for road-paving projects, as ballast for railroad track beds, as poultry grit, and, when finely ground, added to poultry feed to provide calcium for developing eggshells (Churchill 1920; Kochiss 1974). Brooks realized that removal of shell from oyster beds also removed the ideal surface for tiny oysters to attach and grow, so he sought alternative substrates. He found reports from the Connecticut Shell Fish Commission, written in 1882 and 1883, describing methods and materials to collect spat (tiny oysters) in the Poquonock River near Groton. The methods used in Connecticut were very similar to those employed for hundreds of years in Europe and Asia. Bunches of birch branches were stuck in the mud, under 4.3 to 4.6 meters (14 to 15 feet) of water. A single "bush" of 10 centimeters (4 inches) in diameter at the base could generate 25 bushels of oysters, 7 bushels of which were market-size (Brooks [1891] 1996).

In New York's East River and in river estuaries in Connecticut, fishermen began to distribute shell on depleted beds in efforts to collect spat. Laying down shell and other material to collect spat is known as shelling or cultching; the material used as substrate is known as cultch. Oystermen from City Island, Norwalk, and Bridgeport began to save oyster shells, rather than sell them for lime extraction, road projects, or chicken feed. By trial and error, it was found that a clean surface was needed and that the shells should be placed in the water at specific locations and during specific time periods to achieve optimal results. Oyster shells from Rhode Island, which once were considered worthless, were purchased for a song, shipped to New Haven, and spread on the bottom. Gravel sometimes was deposited first to firm up the bottom so the shells wouldn't sink into the mud. After a two-year residency in Long Island Sound, the shells, covered with young oysters, were resold to the original suppliers or moved to other areas where the oysters would grow better (Brooks [1891] 1996).

The practice of cultching has a long history, dating back to ancient times, and remains a very important aspect of oyster propagation efforts when some naturally occurring oysters are present. However, cultching can be a challenge when heaps of shell must be returned to shellfish areas. In large-scale operations, cultch material is power-sprayed over the sides of large vessels, a very expensive operation. In more limited operations, shell often is shoveled from smaller boats under back-breaking conditions. I witnessed a novel, locally engineered cultching operation in Wellfleet on an unseasonably hot day in late June 2006. I was hosting students from a nearby college who were enrolled in a course on field research methods. It was midday and we had just come from a morning of activities at a local salt marsh. The students were hot, muddy, tired, very hungry, and craving seafood, so we went to Mac's, conveniently located next to the pier at Wellfleet Harbor. While we were waiting for our clam chowder, fried clams, and other shellfish delicacies, Andy Koch, Wellfleet shellfish constable, was docking a strange-looking vessel at the Wellfleet pier. It was the town cultch barge, a marvel to behold. The mobile barge has a hole in the middle and the working parts of a sand spreader are secured atop the hole (fig. 1.6). Instead of spreading sand, the barge is rigged to spread surf clam shells in select locations at the bottom of Wellfleet Harbor. As I watched, a dump truck belonging to the Department of Public Works drove up and unloaded the shell onto the cultch barge. With a full load on board, Koch then drove off to lay down the cultch. The clam shells would be the terra firma for the attachment of tiny oyster spat (see plate 1).

Why is cultching necessary and why clam shells? Shells of other oysters are the perfect substrate for small oysters to attach or set. Indeed, that is how oyster bars, rocks, or reefs are formed: new oysters on top of old, layer upon layer. However, oyster shells are not considered to be recyclable material and no deposit is paid for their return to the areas from which the oysters are harvested. When fishermen remove oysters from shellfish beds, the shellfish are sent to local or distant markets. Restaurants or consumers usually are not concerned about the depletion of shell material on the beds. More often than not, the shells go into the dumpster or trash.

Although oyster spat have been know to attach to almost anything, including old shoes, sticks of wood or bamboo, rocks, glass bottles, and tin cans, the optimum set is achieved on material that has the right texture and is free of dirt, slime, debris, algae, and marine organisms. Many substances are used to "catch a set," that is, to attract tiny oysters and encourage their

FIG. 1.6 Wellfleet cultch barge.

attachment. To promote the growth of oysters in natural settings where oysters have been depleted, shells from the surf clam, *Spisula solidissima,* can be used. The shells are readily available from surf clam shucking and processing plants. A surf clam plant in New Bedford is happy to donate its shell refuse to the town of Wellfleet so that the plant doesn't have to deal with shell disposal. The Wellfleet Shellfish Department pays to have the shell trucked to Wellfleet, where it is stored on town property until it is considered clean enough to be used as cultch. During one predetermined week in the summer, it is loaded onto Andy Koch's cultch barge, spread to critical, predetermined sites around Wellfleet Harbor, and becomes available for tiny oysters who are looking for a nice place to settle down.

The timing for the spread of cultch is critical. It takes experience to be able to predict the window of opportunity for the cultch to be successful in catching a set. Water temperature and other environmental conditions must be favorable to oyster reproductive events (spawning). If the cultch is spread too early, it will be covered with algae before the tiny oysters can utilize it. If it is spread too late, the optimal time period for attachment will be missed. It also may be necessary for the cultch to develop a suitable surface-covering of bacteria, which may serve as a spat attractant (Kennedy 1996). If the timing is right and oyster larvae are looking for a home, the cultch will provide an appealing attachment site and another generation of oysters will begin their development into adulthood. So far, Koch has been right on target with respect to the timing of cultch dispersal and Wellfleet Harbor has benefited from several good years of oyster set.

Quahog Culture

Dr. David Belding was a medical doctor and marine biologist who conducted research in the early 1900s on bivalves in southeastern Massachusetts and set up a laboratory on the wharf of the former Chequessett Inn, an upscale resort destination once located in Wellfleet Harbor. The laboratory space was generously provided by Lorenzo Dow Baker, a Wellfleet sea captain who made his fortune by importing bananas and other tropical fruits to the United States from Jamaica in the late 1800s. In Belding's 1912 quahog report to the Massachusetts Commissioners of Fisheries and Game, he noted that native quahogs or hard clams were being overharvested due to excessive demand on beds in New England and New York. He also noted that quahogs were being shipped to the north from southern states to meet the demand for Little Necks, the most desirable size-class of quahog (Belding 1912; [1930] 2004). However, due to the inherent differences in the life history of quahogs compared to oysters, quahog culture was not developed as a means to replenish stocks.

In contrast to the long history of oyster farming, quahog culture is a relatively recent endeavor. Unlike oysters, whose spat are noticeable after they attach to substrates throughout the water column and grow into small oysters, the development of clams takes place under the sediments. Until modern times, very little was known about their reproduction, and clam development was largely a mystery. Some low-tech clam farming was practiced by Native Americans prior to European settlement. The farming basically consisted of moving clams from some areas to other locations where they might be more accessible during certain times of the year. We can think of this as a transplanting program or as a method of wet storage. At times, the stashes of clams were protected from predators by the construction of stockades made from tree branches (Manzi and Castagna 1989; Castagna 2001). No record of a deliberate attempt to culture clams can be found until the mid-1900s.

Shellfish Culture Reaches a Turning Point

The early years of bivalve shellfish aquaculture consisted mostly of moving shellfish from place to place or employing methods to attract spat or seed. Although these techniques temporarily supported the industry, there were several limitations, most notably the necessity for naturally occurring populations capable of reproduction. With continued harvest of the wild shellfish, the reproductive cohort, or "broodstock," also were becoming depleted. It was time for the industry to develop a new way to propagate future generations of bivalves.

Chapter 2

Birth in the Laboratory

In the late nineteenth century, while engaged in research on oyster biology, William K. Brooks demonstrated the possibility of producing oysters in the laboratory. He succeeded in obtaining eggs from female oysters and sperm from male oysters and promoting fertilization in glass dishes. He realized that great economic potential could be tapped by producing vast numbers of oysters in the laboratory and growing them to a point at which they would be more likely to survive in the wild. His embryonic oysters developed into larvae (an early developmental form) but neither Brooks, nor Julius Nelson, another biologist who was conducting oyster research in New Jersey, were able to keep the larvae alive. Despite the early accomplishments in achieving fertilization in the laboratory, the potential for artificial rearing of oysters was not realized until half a century later.

Meeting the Growth Requirements of Larval Shellfish

William Firth Wells discovered a method to culture various types of bivalve molluscs such as oysters, quahogs, bay scallops, surf clams, softshell clams, and mussels, while he was experimenting with a clarifier, a novel centrifugation device that had been designed to separate components of milk based on density. In the 1920s, while working in West Sayville, New York, Wells applied the device in efforts to separate the components of seawater. He was able to enrich seawater preparations for the larval forms of bivalves and other fractions of the same preparations with phytoplankton, the algae on which shellfish larvae feed. He had some success growing the larvae and received a patent on his method in 1933 (Manzi and Castagna 1989). As promising as the technology may have been, it was not readily adapted by commercial shellfisherman.

The next chapter on bivalve culture took place in Milford, Connecticut, during the mid-1940s and 1950s, when Drs. Victor Loosanoff and Harry C. Davis worked for the U.S. Fish and Wildlife Service at the Laboratory of the Bureau of Commercial Fisheries (now known as the National Marine Fisheries Service). Their experiments led to increasingly productive methods for the culture of commercially important bivalves. Loosanoff and Davis also experimented with different unicellular algal species and identified several varieties that filled the dietary requirements of tiny oysters. The use of specific strains of phytoplankton and development of techniques for their culture became known as the "Milford method." Similar progress was made in Virginia by Dr. Michael Castagna, who used running, unfiltered seawater (Carriker 2004). Adopting these methodologies, several small hatcheries emerged. Most were successful in producing larvae but were faced with challenges in providing sufficient food for larval growth and development and in preventing mortality due to bacterial contamination. Further challenges arose when the small shellfish were placed out on the beds. After all the effort of producing the tiny shellfish, almost all of them were devoured by predators when they were no longer in the protective custody of the hatchery. One of the pioneer hatcheries, founded in Virginia, had sporadic success using refinements that would not be approved today, such as controlling predators with the use of fish carcasses laced with agricultural pesticides (Manzi and Castagna 1989).

In 1959, Joseph Glancy was hired by the Bluepoints Oyster Company in West Sayville, New York. Earlier in his career, Glancy knew and worked with Wells and had invested in his own shellfish leases in Great South Bay. He also worked on clam culture as an extension of his involvement with the oyster business. Glancy expanded on Wells's ideas and devised his own improvement to the rearing of larval bivalves. Two of the main problems with bivalve culture in that era were the selection of appropriate phytoplankton and the lack of methods to provide food in sufficient quantities. While expanding his warehouse space, he created a type of greenhouse area in the hatchery using roofing material made of clear plastic sheets. The "roof" allowed the penetration of sunlight and greatly enhanced the growth of phytoplankton, thus causing an algal "bloom" in his filtered, clarified seawater from which large particles and organisms previously had been removed. This major innovation for growing phytoplankton, patented in 1965, became known as the Glancy method (Manzi and Castagna 1989; Gosling 2003). The combined larval rearing strategy of using natural seawater and forced algal blooms is referred to as the Wells-Glancy method and is still used as a blueprint for bivalve culture in the Northeast.

Other hatcheries devised further refinements to the methodology. Paul and Matoria Chanley, working at the Plock hatchery in Greenport, New

York, devised a method to sterilize seawater chemically, using laundry bleach. This was an important step in hatchery operations because it addressed the serious problem of bacterial contamination, which could quickly destroy all the larvae in a rearing tank. Although the Plock hatchery seemed to solve the bacteria problem, like others, the company did not have success when the juvenile clams were planted in the field (Manzi and Castagna 1989; Castagna 2001).

With improvements to hatchery operations and the possibility that this strategy for growing shellfish could be commercially successful, two Long Island oyster companies invested in bivalve culture. Frank M. Flower and Sons Oyster Company in Oyster Bay and Long Island Oyster Farms, Inc., in Northport tried to manipulate the water supply in which larvae were grown. Flower and Sons used filtered seawater with low salinity that was free of any living organisms, while Long Island Oyster Farms used the power-generator cooling lagoon of its neighbor, the Long Island Electric Company. In Massachusetts, ARC (Aquaculture Research Corporation) in Dennis became a pioneer in hatchery operations when the company achieved economic success in rearing larval bivalves to planting size. ARC and other companies that had success in their hatchery operations opened the door for the expansion of bivalve aquaculture in the Northeast when they demonstrated the ability to reliably supply clam and oyster seed to culturists.

University Research Leads to Improvements in Culture Methods

Aside from their value as food, bivalve shellfish are important inhabitants of coastal ecosystems, where their ability to filter tremendous volumes of water has been recognized for its critical role in the environment. University scientists have been involved in many aspects of shellfish aquaculture, not only to buoy the industry, but also to lead efforts to restore shellfish to locations where they were once abundant. They have improved and standardized hatchery and culture methods, initiated selective breeding programs, and spearheaded shellfish restoration efforts. The Haskin Shellfish Laboratory at Rutgers University had its humble beginnings in the late 1800s and early 1900s and included a floating laboratory, the barge *Cynthia*. Julius Nelson, a former student of William K. Brooks, was in charge of the program and a pioneer in the study of the ecology of oyster larvae. His sons Thurlow Nelson and J. Richards Nelson and grandson Richard followed in his footsteps and continued the family traditions in oyster research (Carriker 2004). Several eminent shellfish researchers later trained under the Nelsons, including Dr. Harold Haskin, for whom the lab later was named.

Scientists at the College of William and Mary's Virginia Institute of Marine Sciences were, and continue to be, important contributors to refinements in bivalve culture and shellfish restoration projects. The College of William and Mary helped growers by providing some of the first manuals and guidelines, offering courses to train those interested in hatchery operations, and proposing suggestions regarding field-growing conditions, still a bottleneck in many aquacuture operations (Dupuy, Windsor, and Sutton 1977). Hatcheries cannot afford to raise bivalves that are larger than a few millimeters (tenths of an inch) in size because of the large amount of food that must be provided. It was discovered that a middle step, a nursery, is important for the success of farming operations because of the tremendous mortality when tiny seed bivalves are placed in the field.

In the early 1950s, a marine biology program was established at the University of Delaware that continues to contribute to research in shellfish aquaculture. In the Northeast, research in aquaculture methodology and applications as well as shellfish restoration is being conducted at many institutions. The State University of New York at Stony Brook and Cornell University contribute to shellfish aquaculture research, restoration, and education in Great South Bay and Peconic bays. The University of Connecticut and scientists at the National Marine Fisheries Laboratory in Milford, Connecticut, maintain a visible and healthy shellfish aquaculture research program. In Rhode Island, two institutions, Roger Williams University and the University of Rhode Island, work together to define and tackle shellfish aquaculture research priorities for Rhode Island. In Massachusetts, the Marine Biological Laboratory (MBL) at Woods Hole and the Massachusetts Institute of Technology, through their SeaGrant programs, have made many contributions toward aquaculture. The Boston University Marine Program (BUMP) and the University of Massachusetts also have research programs with shellfish aquaculture components. New Hampshire, with only seventeen miles of coastline, is represented by the University of New Hampshire, where a fledgling research program has been initiated, while in Maine, shellfish aquaculture research programs are in place at the University of Maine and at the University of New England. Some of these university programs receive support from federal sources such as NOAA (National Oceanic and Atmospheric Administration) and from organizations such as the Nature Conservancy. The discoveries and improvements to shellfish culture operations made by scientists not only benefit the commercial growers, but are also important to environmental and conservation efforts to return shellfish to areas from which they have disappeared and to enhance declining populations in order restore the role of these organisms in estuarine and coastal habitats.

Current shellfish hatchery practices evolved from research conducted at universities and funded by federal, state, and nonprofit organizations. These partnerships remain important for the present and future well-being of the industry, especially as wild stocks decline.

Chapter 3

Bivalve Shellfish of Commercial Importance

Bivalve shellfish are molluscs, a group of animals (Kingdom: Animalia) with soft bodies. The name is derived from the Latin word for "soft." In the United States, the group name is often spelled "mollusk." Phylum Mollusca is a large grouping of animals that also includes snails, as well as some varieties without shells, such as squid. The term "shellfish" is used to include all shelled molluscs. The class Bivalvia encompasses the molluscs with two valves or shells, many of which are candidates for shellfish farming. Each mollusc also belongs to a specific order and family, but the most important identifier is the scientific name for its genus and species (see table 3.1).

Oysters

Oysters are epifaunal creatures, that is, they must live above the sediments, a significant difference from the residential area used by quahogs and other clam varieties. This residential living space difference also will be reflected in adaptations in their anatomy and physiology. Much has been written about the biology of oysters, with detailed information in the seminal works of Brooks ([1891] 1999) and Belding (1912; [1930] 2004), who both addressed the need for artificial culture of oyster to supplement declining wild resources. Brooks chronicled the noticeable declines in the Chesapeake Bay oyster fisheries while Belding focused his observations and recommendations to conditions on Cape Cod and other areas in New England. Galtsoff (1964) added to our knowledge of bivalve biology by providing descriptions of physiological processes along with detailed anatomical drawings that are still consulted by scientists and used as reference for current studies. *The Eastern Oyster* (Kennedy, Newell, and Eble 1996) is a collection of chapters written by eminent scientists that reflect the current state of knowledge of oyster biology, reproduction, and ecology. Kurlansky (2006) describes the influence of the oyster on the history of New York

Table B.1
Bivalves of Commercial Importance in the Northeast

Common name	Genus	Species
Eastern oyster	*Crassostrea*	*virginica*
Northern quahog	*Mercenaria*	*mercenaria*
Softshell clam	*Mya*	*arenaria*
Sea scallop	*Plactopecten*	*magellanicus*
Bay scallop	*Argopecten*	*irradiens*
Blue mussel	*Mytilus*	*edulis*
Razor clam	*Ensis*	*directus*
Surf clam	*Spisula*	*solidissima*
Ocean quahog	*Arctica*	*islandica*

City in a fascinating account that explains how that city became the oyster capital of the world.

The Eastern oyster, *Crassostrea virginica,* is sometimes called the Atlantic oyster or the American oyster. The latter name would imply that no other species of oyster is native to the United States, which is not the case. The Olympia oyster (*Ostrea conchaphilia*) is native to the West Coast, and before the introduction of other oyster species was the predominant oyster in California, Oregon, and Washington. Eastern oyster is the preferred name for *C. virginica,* which is distributed along the Western Atlantic coast from Canada to Florida, along the Gulf of Mexico, and in the Caribbean, Brazil, and Argentina (Carriker and Gaffney 1996). Oysters are easily identifiable because of their rough and irregularly shaped shells (plate 2). Despite the fact that they are grown regionally and marketed with specific trade names such as Wellfleets, Cotuits, Blue Points, Plymouth Rocks, and others, they all belong to the same species. This seems to contradict the observation by growers that their oysters have a distinctive shape, shell coloration, taste, or shell thickness. For the most part, these differences are due to the growing conditions, not to the species of oyster.

The European flat oyster, *Ostrea edulis,* is common in the Mediterranean and Eastern Europe, where it has been cultivated since antiquity. Once considered an aphrodisiac, its name means "edible oyster," reflecting the fact that it also has been valued as a food source. This species has been introduced to both the east and west coasts of the United States and to Japan.

Northern Quahog

The hard clam, *Mercenaria mercenaria,* is known by several names, including hard-shell clam, round clam, quahog (also spelled quahoag), mud clam, and northern quahog (plate 3). The latter is the scientifically accepted common name for the species. The term "quahog" stems from the shortening of the Native American name for these clams, the Narragansett *poquauhock* or Pequot *p'quaghhaug* which means hard shell. Most New Englanders pronounce the word as "co-hog," but Block Islanders and western Rhode Islanders often say "kwa-hog."

Similar to many other marine species, the name and biological classification for the northern quahog have undergone revision. The name "mercenaria," from the Latin term for money, was given to the clam by a famous biologist, Carl Linnaeus, who developed the currently used system of biological taxonomy, the naming and classification of living things. He coined the quahog's official name in 1758 after becoming acquainted with its use for the production of wampum (described in chapter 4). Biologists grouped this clam with others in the genus *Venus* and used the name *Venus mercenaria* until 1936, when the quahog was recognized as a separate genus (Kraeuter and Castagna 2001).

The quahog sometimes is named according to the size of the clam and its market value. Littlenecks are the smallest and most valuable clam that can be legally harvested and sold (1–2 inches [2.5–5 cm] in width and 400/bushel). Cherrystones are medium-sized (3–3.5 inches [7.7–8.9 cm] wide, 150/bushel), and chowders are the largest clams (over 3.5 inches [8.9 cm] wide; 100/bushel). Terms such as topnecks or topcherries (2–3 inches [5–7.7 cm] in width; 200/bushel) are sometimes used to further divide clams in the intermediate size ranges (Getchis 2006). Natural beds of quahogs in the mid-Atlantic were the source of clams used in many clam products such as chowders and soups, processed by Borden Food Corporation and Campbell Soup Company. When natural hard clam beds became depleted, large companies turned to other species such as surf clams and ocean quahogs.

The term northern quahog is a bit misleading; although *Mercenaria mercenaria* is found in the northern United States and as far north as the Gulf of St. Lawrence, it also is found along the entire Atlantic coast of the United States and its distribution is reported to extend through the Gulf of Mexico. Scientists disagree whether some Gulf Coast clams are really *M. mercenaria;* they may be a close relative, *Mercenaria campechiensis,* the southern quahog, whose range overlaps that of *M. mercenaria* in the south-eastern United States.

Quahogs are long-lived and can survive perhaps forty years or more. Similar to most clams, they live below the sediments and are thus categorized as an infaunal species.

Anyone who wants to know the detailed features of northern quahog biology, taxonomy, growth, reproduction, shell structure and formation, detailed anatomy, reproduction, development, genetics, and physiology can consult a number of superb sources. In 1982, McHugh and colleagues compiled a summary of 2,233 reports about *Mercenaria* in the scientific literature and an additional 460 titles appeared before 1988 (McHugh and Sumner 1988). David Belding's landmark studies covered the biology of quahogs as well as oysters (Belding 1912; [1930] 2004). An overview of clam biology, *The Northern Quahog: The Biology of* Mercenaria mercenaria, written by Rice, appeared in 1992. In 2001, all known scientific information about hard clams was compiled by Kraeuter and Castagna in a volume that serves as the handbook of clam information for the scientist (2001). I summarize and refer to the most important aspects of clam biology for those who want to know the simple, basic facts about clams, but refer the reader to the above sources for further details and scientific reports.

Softshell Clam

Mya arenaria is the biological name for the softshell clam, whose range extends along the Atlantic from subartic regions to North Carolina (plate 4). The scientific name roughly translates to "shellfish that lives in the sand." In the early 1800s, this Atlantic species was introduced to the West Coast, where it was harvested commercially in the San Francisco Bay area. Today, its West Coast numbers have declined to a point where it is only taken for bait or recreation (Abraham and Dillon 1986). The softshell clam is still a very important fisheries species. Although it once was abundant in the Northeast, its fate has been similar to that of other bivalve shellfish and its numbers have declined.

The softshell clam is found close to shore, where it burrows deep into the sediments with its foot. Its preferred habitat, in the intertidal areas of estuaries and bays, makes it particularly sensitive to pollution. Many beds in the Northeast have been declared uncertified for harvest of softshells due to polluted waters. Some beds have been destroyed completely by a form of clam cancer that may persist at low levels in Northeast softshell clam populations.

Softshell clams are also known as Ipswich clams, after the town in Massachusetts that once had a thriving softshell clam industry. Other familiar names include long-necked clams, because of its long siphon (described in chapter 5), gaper, because the shell does not close completely around the body (an anatomical condition that is responsible for its short shelf-life), and squirt clam or piss clam, because it emits a stream of water when stressed, an observation that can be authenticated by anyone who walks over a tidal flat where these clams may be buried. Softshell clams are not eaten raw but are the clam of choice for steamers, the familiar dish of steamed clams served with melted butter. As their name implies, softshell clams have a soft, brittle shell, and care must be taken in the methods used for their harvest. This clam, which has a life span of at least twelve years in the wild, is still commercially valuable and is currently the focus of some limited aquaculture operations. Information about the biology of softshells can be found in Belding's reports (1915; [1931] 2004) and in a U.S. Fisheries and Wildlife Service species profile prepared by Abraham and Dillon (1986).

Scallops

"Scallop" is the answer to an old riddle that asks, "What can clap without hands, see but has no head, and swim without arms, legs or tail?" The word "scallop" is derived from the ancient Gothic word SKAL which apparently referred to any type of hard covering, including nut shells and even roofing material. In France, the word evolved to "escalope" and later became specialized to indicate the particular type of shell with a very distinctive, wavy border. The word also was used to indicate any curvy edging or design that resembled the border of the shell (Cox 1957). The word "pecten," part of the genus name for many species of scallop, is from the Latin term for comb, referring to the ridged surface of the shell that resembles a hair comb.

The scallop shape is so pleasing and symmetric that the design has been used in painting, sculpture, and carvings from classical times. Its use was expanded when the motif was incorporated into coins, metal works, furniture, and architectural details. During medieval times, the scallop often was used as a symbol in heraldry and it appeared in very stylized forms on coats of arms and also embossed on the armor itself as early as the mid-thirteenth century.

The scallop also had deep religious symbolism for pilgrims during the Middle Ages. Pilgrims traveling to the shrine of Saint James the Great in Santiago de Compostela, Spain, often adorned their clothing with scallop symbols, the saint's emblem. It is believed that the pilgrims would carry a scallop shell and that when they depended on the kindness of folks along their route for food or drink, they would take only as much as would fit into the shell. The reason for the association of the scallop shell with Saint James is not clear, but his influence is readily seen in the name of one of the most familiar and popular scallop dishes, Coquille Saint Jacques, a preparation in a creamy, buttery sauce described in many recipe variations (see appendix for a simplified version).

Perhaps the most famous depictions of a scallop shell are in various artworks that portray the legend of the birth of Aphrodite. The Greek goddess emerges from the sea behind or upon a large scallop shell. In the 370 to 360 B.C. *Birth of Aphrodite,* the goddess is seen demurely peeking out from behind an upright scallop shell. Aphrodite became the goddess Venus in Roman mythology. *The Birth of Venus,* painted by Botticelli in about 1485, depicts a naked Venus rising from the sea atop a single valve of the scallop shell. The painting is sometimes referred to, with tongue in cheek, as "Venus on the half shell." Today, the scallop design is used in several company logos and various symbols and designs, perhaps the most familiar of which is the multinational company, Shell Oil.

In the Northeast, two types of scallops are of commercial importance. The sea scallop, *Plactopecten magellanicus,* is the larger variety and is found in deep water from Newfoundland to North Carolina (plate 5). It is also called the giant scallop, smooth scallop, deep sea scallop, or ocean scallop (Naidu and Robert 2006). Dredging is the usual method of harvest. Scallops must be shucked on board the dredging vessels because they can't completely close their shells and have a short shelf-life out of water. The large adductor muscle (20 to 40 scallops per pound) is the product available to U.S. consumers.

The smaller bay scallop, *Argopecten irradiens,* is found closer to shore from New England to the Gulf of Mexico. Its species name, *irradiens,* refers to the raised ribs on the shell surface that extend outward like rays (plate 6). Similar to sea scallops, the product is the adductor muscle (50 to 90 scallops per pound), obtained after the bay scallops are shucked. Many of the bay scallop products currently sold in the United States are frozen adductor muscles from scallops cultivated in China. The bay scallop was introduced to China in the late 1980s, where it has been a very profitable specimen for aquaculture operations that utilize suspension nets in the water column.

Unlike oysters and clams that tolerate some time out of the water, scallops prefer to be submerged at all times and do not fare well in interidal locations. The bay scallop is a species of interest for aquaculture in subtidal locations in the United States, where it is being cultured for restoration purposes and to test the feasibility of the species as an aquaculture product. In contrast to sea scallops, which can live for twenty years, bay scallops have a very short life span. They generally don't survive longer than two years, and they can't be harvested legally in the wild until they are in their second year. In order for culturists to consider growing bay scallops, it will be important to find ways to grow them to appropriate, market size during their first year or to discover techniques that ensure good survival during their first winter. Sea scallop aquaculture has been attempted since the 1970s but has proved labor intensive and may not be economically feasible (Parsons and Robinson 2006).

Wellfleet physician David Belding also studied the biology and reproduction of bay scallops during his tenure as biologist in the Massachusetts Department of Fisheries and Game (Belding 1910; [1931] 2004). A more recent report on bay scallops was published as a species profile by the U.S. Fish and Wildlife Service (Fay, Neves, and Pardue 1983). Sea scallops are the subject of several technical reports, including an Essential Fish Habitat Source Document (Hart and Chute 2004). A description of the biology and reproduction of many commercially important scallops and a global perspective on the scallop industry has been collected and edited by Shumway and Parsons (2006).

Blue Mussels

The blue mussel, *Mytilus edulis* (edible sea mussel) has an oblong to pear-shaped shell, shiny, dark blue and bordering on black on the outside with a violet or white pearly layer on the inner surface (plate 7). Wild blue mussels commonly are found in large beds on rocky intertidal or subtidal areas and they also attach to rocks, pilings, jetties, piers, and other natural and man-made structures. They live for about 18 to 24 years (Gosling 1992). The attachment of mussels to each other and to structures is mediated by long, brown, tough fibers known as byssal threads, commonly referred to as the "beard."

Blue mussels are found worldwide from the north polar regions to temperate waters. South Carolina is the southernmost range in the western Atlantic. For many years, the state of Maine had the largest commercial U.S. fishery for wild blue mussels. However, since the 1980s, aquaculture has supplied increasing production of blue mussels in Maine. Today, the largest source of cultivated mussels sold in the United States is from Atlantic Canada, near the Gulf of St. Lawrence: New Brunswick, Nova Scotia, and Prince Edward Island. The industry began in 1981 with the introduction of culture techniques on long-lines in deep water, and blue mussels are now the most important bivalves cultured in this region.

In contrast to farmed oysters and little neck clams, which usually are eaten raw, mussels are prepared by a number of cooking methods. They can be simply steamed in water or with some wine to produce the popular dish moules mariniere (see appendix for a recipe). Those who consume mussels will notice that some of the meats are pale yellow or creamy white while other are yellow to orange in color. The paler meats are those of males and immature females while brightly colored meats are those of mature females. Gosling (1992) has edited a comprehensive scientific volume that summarizes the biology, evolution, ecology, and aquaculture of mussels.

Razor Clams

The razor clam, *Ensis directus,* is also known as the jackknife clam because of its shape. Its Latin name translates as "straight sword." The adult razor clam is 16 to 17 cm (about 10 inches) long and thin with a very slight curve and an overall shape that resembles an old-fashioned straight-edge razor within its protective sheath (plate 8). The name "razor" might seem appropriate to those who have walked the flats barefoot and have had their foot sliced by the very sharp shell of this species. This clam can burrow deeply into the sediment and is the fastest-digging clam in the Northeast, easily able to elude a shovel or rake as it extends its foot to lengths as long as its body, expands the ends of the foot to produce an anchor-type structure, then instantly draws the rest of its body to the new location. In addition to the action of the muscular foot, the razor also squirts water ahead of the path of the foot to soften or clear away mud to expedite its transit in the sediments.

The razor clam also has been called the razor fish, a reflection of the swimming ability of the species. The clam can move through water in quick, jerky, backward motions as it emits jets of water. The fact that the razor clam will not stay put is a significant challenge for those who might consider it for its commercial value as an aquaculture species. Although the razor clam can be used in most types of clam recipes, including soups and stews and for frying, it is difficult to harvest from the wild and difficult to contain on shellfish farms.

Razor clams can live up to five years, forming extensive beds that can markedly alter the characteristics of the sediment. This clam has invaded Europe and has colonized in the North Sea. It was first found in the Elbe estuary in the 1980s, most likely carried there as tiny larvae in ballast water.

Surf Clams

Surf clams, (*Spisula solidissima*), also called hens, sea clams, and giant clams, are large clams that are found in deep water from the Gulf of Maine to North Carolina. Their scientific name implies that their shell is dense and solid, and they are often seen along ocean beaches where they are gathered up to be used as ash trays and soap dishes (plate 9). The usual method of harvest is dredging, but some areas have permitted only hand harvest by diving. These commercially valuable clams once were used mostly for bait but are now used in cooked clam dishes and in products such as clam chowders and clam pies. They are commonly employed to prepare the clam chowder stock that is used in many restaurants as the staring point for regional or restaurant-specific clam chowder variations. There is current interest and potential for employing *Spisula* for aquaculture applications, where they might be harvested at smaller sizes.

Ocean Quahogs

Ocean quahogs, *Arctica islandica,* are found over a wide range in the North Atlantic. The clams were first classified as *Venus islandica* by Linnaeus in 1767 but reclassified in 1817 in the genus *Arctica.* This is a cold-water clam that is found in water temperatures from 0 to 19°C (32 to 66°F) and spawns (releases eggs and sperm) at about 15°C (59°F). It has been called a number of other names, including ocean clam, black clam, mahogany clam, and mahogany quahog. The dark shell color of adults is due to iron complexes that are deposited in its outer shell layer, the periostracum. Because of its short siphons, it is found under very shallow sediment layers.

The ocean quahog is the Methuselah among bivalves, with the longest lifetime of all the commercial bivalves. Clams over 100 years old have been observed and 200-year-old specimens have been reported. An 8.6-cm (3.4-inch) specimen, dredged off the Icelandic coast in 2007, was estimated to be between 405 and 410 years old! The clam was nicknamed "Ming" after the Chinese dynasty during which it was but a larva. The clam grows throughout its life, but its growth rate decreases dramatically after it reaches twenty years old.

Ocean quahogs are found at depths of 8 to 256 meters (26 to 840 feet). They are harvested by hydraulic dredges and have not been cultured. In Maine, the harvest consists of clams that are about 50 mm (2 inches) in length and sold in the shell. Larger ocean quahogs, which are found in the Middle Atlantic, are used for processed clam products (Jacobsen and Weinberg 2006). Although this clam is currently not an aquaculture candidate, there has been increased harvest pressure on the species, partially to make up for declining surf clam stocks.

From the Natural World to the Farm

Many bivalve species of commercial importance live in the Northeast, but whether a species is amenable to culture and farming depends on a number of factors, including the ability of culturists to foster reproduction in the laboratory and to develop conditions and equipment to ensure the survival and growth of small, young specimens. In the following chapters, we will explore the biology and reproduction of these potential marine crops and the methods that have been employed to propagate them and enhance their survival.

Chapter 4

You Can Judge a
Bivalve by Its Cover

Although some shellfish aquaculturists are able to harvest their crop on a year-round basis, farmers who have pitted their oysters or whose quahog farms are covered with thick layers of ice may experience some down time or finally have the opportunity to take a vacation in winter (fig. 4.1). Others remain busy as they ply different trades or continue with ongoing commitments such as part-time jobs and winter-term college courses. Some shellfish farmers work in the construction business and can pick up extra work during winter. It's helpful to have a backup source of income to hedge against unpredictable problems such as damage of equipment due to weather or loss of crops due to disease. Part-time, small-scale shellfish farmers who are working on the flats to supplement their income often have other responsibilities.

Outdoor activities may grind to a halt during the deep freeze of winter, but this is one of the busiest times for indoor shellfish cultivation at hatcheries as they gear up to mass-produce seed oysters, clams, and scallops for commercial sale or for restoration projects. Hatcheries maintain a critical schedule so that the timing of their production of seed bivalves will coincide with needs of the growers.

The winter of 2006 was a bit unusual in the Northeast. The season started out with very mild temperatures. By late January, there was very little snow on the ground but winter finally arrived when a cold front from Canada, which New Englanders call a Montreal Express, descended its reach to southern New England and New York. Biting winds sent the temperatures into the single digits but the cold snap was not entirely unwelcome. Shellfish aquaculturists appreciate a winter deep-freeze because some shellfish pathogens are sensitive to extreme cold and may not survive the winter in the environment or in their hosts. Nevertheless, under the very cold conditions of the 2006 winter, some shellfish farmers were anxious. They had not ushered all their animals and equipment to safety before thick ice sheets blanketed the creeks, coves, harbors, and near-shore

FIG. 4.1 Winter ice over shellfish farms in Wellfleet, Massachusetts. *Photo: Captain Andrew Cummings*

areas. For those unlucky aquaculturists, a considerable amount of damage was done to their shellfish and their gear was strewn about for miles.

During a Northeast winter, shellfish become dormant. They close up, do not feed, and do not grow. Shell enlargement is halted and physiological processes are in standby mode. The anatomy and physiology of these molluscs provide clues to understand how these organisms will regain function when the water begins to warm. To learn about the adaptations and characteristics that contribute to their ability to be cultured and farmed, we will examine the bivalves, starting from the outside in this chapter and moving in.

The Shell

Current efforts to grow and harvest bivalve molluscs such as oysters, scallops, and various species of clam are rooted in the commercial value of shellfish as food items. While it cannot be disputed that bivalve molluscs have been valuable historically for this purpose, human interest and fascination with them is not limited to their effect on the palate. From the human perspective, the most distinctive and fascinating feature of all bivalve molluscs is their shell. The shell itself, and other creations fashioned from shell material, have had value as human ornamentation and symbolic communication since prehistory. The finding of marine shell beads in archeological excavations in West Asia and Africa indicates that humans may have fashioned shell jewelry as long ago as 100,000 years (Vanhaeren et al. 2006).

Although the shell has ornamental value for us, it plays a critical role for the mollusc. Whether it is colorful or drab, smooth or rough, shiny or dull, the shell is the major line of defense against predators. The two valves of the shell serve as protective armor and as sites for attachment of internal structures. For burrowing clams, the sleek design of the shell also protects the internal organs from sand and mud. Although all bivalve molluscs have shells, the shape, coloration, articulation, and design differ in each species; coloration sometimes varies even within a species. Indeed, the shape and patterns of shells are identifying features for taxonomists who classify molluscs, the biologists who look for the evolutionary relationships among them, and conchologists who collect and study shells.

Many of the Northeast states have adopted "state shells." In 1987, Rhode Island named the northern quahog as its official state shell, while Massachusetts picked the wrinkle whelk, *Neptunea lyrata decemcostata*. New York followed suit by naming the bay scallop, *Argopecten irradiens* in 1988, and, in 1989, Connecticut followed the lead of Mississippi and Virginia and gave the honor to the eastern oyster, *Crassostrea virginica*.

The size of a bivalve is measured by its shell, and is often an important parameter in determining whether the specimen can be legally harvested (fig. 4.2). A key reference point on the outer shell is the umbo or beak, the oldest part of the shell and the area to which new shell is deposited during the growth of the bivalve from its days as a larva. The animal also has an anterior side, near its mouth, and a posterior side, which, in the case of clams, is near its siphons, the tubes used for water intake and output. The placement of the siphons at the posterior end of the clam is an evolutionary adaptation that allows the animal to burrow into sediments head first, while still allowing for water intake and feeding activity. The deeper a clam burrows, the longer its siphons must be. The dorsal side of clams and scallops is at the hinge, while the ventral side is at the shell margin. The hinge of the oyster is actually at its anterior side, so its dorsal and ventral surfaces are oriented differently from a clam's. A bivalve's height is the distance from the hinge region to the outer margin of the shell and is the most important measurement applied to oysters (fig. 4.2). The length of the bivalve, a measurement used for softshell clams, is its largest measurement at 90 degrees from its height. The width of the bivalve, an important measurement for commercial quahogs, is its thickest part and includes the measurement of both valves. Mussels and clams have paired valves that are the same size and

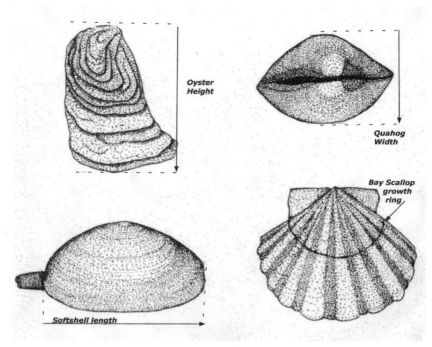

Oyster
Height

Quahog
Width

Bay Scallop
growth
ring

Softshell length

FIG. 4.2 How to measure shellfish. Solid lines refer to measurements used to determine if the bivalve can be legally harvested. *Drawing: Marisa Picariello*

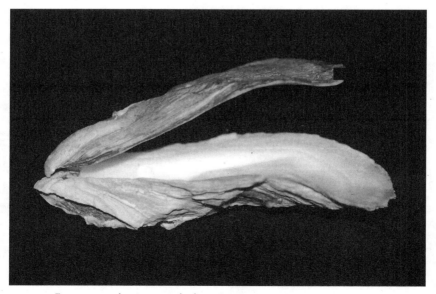

FIG. 4.3 Eastern oyster showing cupped valve. *Photo: Nina Picariello*

depth. In contrast, oysters and scallops have valves that are not exactly identical. Valves also can be designated as right and left and can be assigned when the viewer directly faces the posterior end of the mollusc. The valve on the viewer's left will be the left valve and the one on the right will be the right valve. One valve, the left in oysters, is more deeply cupped, surrounds the soft body of the mollusc, and is cemented to the substrate, while the opposite valve fits like a lid on top of the shell (Fig. 4.3). The deeper valve— on the right side of the bay scallop and on the left side of the sea scallop— forms an indentation into the sediment while the complementary valve serves as the cap.

Shell Composition

The main component of a bivalve mollusc shell is the mineral calcium carbonate ($CaCO_3$), which is embedded within a matrix of a protein known as conchiolin. The shell is composed of three layers:

1. The outer layer is the periostracum. It is made up of conchiolin and is very thin. When molluscs are harvested, the periostracum might not be seen because it is often abraded. The best chance of seeing it would be on a surf clam shell that has been deposited recently on the beach. The periostracum often appears as a brown, flaky covering on the outside of the shell (plate 9). It is not mineralized and is believed to protect the mineralized layers from being dissolved by seawater.

2. The middle shell layer is the prismatic layer. It is made up of different crystalline forms of calcium carbonate: aragonite and calcite.

3. The inner shell layer is the nacreous layer. It can be matte or shiny and is iridescent in some species, where it forms the characteristic "mother of pearl." The pallial line, seen inside empty shells, is actually a scar that marks the former attachment of the mantle to the shell (plate 9).

Shell Growth

The growth of bivalves usually is measured by examining growth patterns of the shell. The outward growth of the shell from the hinge area is accompanied by an increase in thickness. Because growth is occurring in several dimensions, fast growers may have thinner shells than more slowly growing animals. Growth is not uniform throughout the year; during winter dormancy, growth will cease. This period often leaves a telltale mark in the

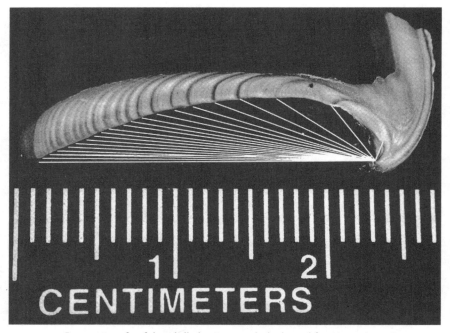

FIG. 4.4 Cross section of surf clam shell, showing growth checks used for aging. *Photo: Adriana Picariello*

shell known as a growth check. Similar to tree rings, growth checks often can be used to determine the age of a specimen. Markings similar to growth checks also may be caused by suspensions in growth due to disease, predation, handling, and other stressors (Gosling 2003). While the growth check in the bay scallop is fairly easy to discern visually or by running one's finger along the shell and feeling a ridge (see fig. 4.2), and one can sometimes assess the growth of clams by visual inspection, the growth checks in oysters and mussels are almost impossible to read. When it is important to know the age of specific shellfish, scientists prepare cross sections of the shells to make the growth checks visible (fig. 4.4).

Coloration

Coloration of the shell is due to pigments whose production is under genetic control. Environment also will affect the expression of color. Oyster shells are usually white or creamy, sometimes with brown or grey markings (plates 2 and 10). I have seen some that are almost black. The inside of the shell is white with a shiny, pearly glaze. The color of blue mussels depends on their age and location (plate 7). In the intertidal, *Mytilus edulis* are dark blue or blue-black; in the sublittoral, when they are constantly under water, their shells are thinner and brownish, although they may have some dark blue radial markings (Gosling 2003). The softshell clam, which can reach up to 15 cm (5.9 inch) in length, has a white, elliptical shell that, in some specimens, is partially or fully covered by a thin, brownish periostracum (plate 4).

Bay scallop shells have a distinctive shape and appearance and exhibit a variety of colors (plate 6). The wing-type structures on either side of the hinge are known as auricles. The bay scallop shell can grow up to 9 cm (3.6 inches) in height and is rarely of uniform color. Its deeply grooved shell, with 13 to 22 ribs radiating from the umbo, can vary from white to brownish grey with numerous color permutations and markings. The valves of the sea scallop, which can reach a height of over 8 cm (3.1 inches) and lengths of up to 17 cm (6.7 inches) in adults, vary in coloration; the lower valve is white, flat, and smooth, while the upper valve is light to pale brown, sometimes with pinkish tints; it has a slight convex shape and very fine ribs (plate 5). Scallops also exhibit deep concentric rings over the ribbed surface, which represent annual growth rings.

The most common form of the northern quahog is *Mercenaria mercenaria alba*. It has a whitish shell with deep concentric grooves. Freshly harvested quahogs appear dark blue or indigo when they are removed from the sediments but the color quickly fades during dry storage to reveal the white shell. About 3 percent of quahogs found in the wild have brownish zig-zag markings, sometimes called Indian blanket designs or "W" patterns (plate 11). These variants are *M. mercenaria notata* and are hybrids of the *alba* variety and another dominant genotype that produces a dark-colored shell. The dark-colored variety is rarely seen in the wild but can be observed in hatcheries when *notata* are bred. The rim of the inner surface of the quahog shell often has a distinctive, deep purple color and was the source of wampum (plate 12). Wampum has played an important role in the interactions between Native American tribes and the history of colonial settlement in the Northeast. To understand its contribution to United States history, it is important to appreciate the quahog shell.

Wampum

Mollusc shells, particularly those of quahogs, always have had importance in Native American cultures. They initially were employed as tools for a variety of personal and practical purposes. Two quahog shells were used as tweezers to pluck facial hair. Individual quahog shells were used as scrapers in woodworking to form canoes and bowls and also as primitive hoes for digging. Some shells were crushed and embedded in pottery. In North America, archeological findings of rounded, smooth pieces of purple quahog shell in graves that date back over 4,000 years suggest that these shell pieces were kept as charms or talismans. The Algonquins called them *suckauaskeequash* or blackeyes. Small shells or pieces of shell were pierced and could be strung on hemp or twine made from fibers produced from local plants. The resulting beads had a variety of shapes.

Tube-shaped beads made their appearance as early as A.D. 200 and were fashioned with stone tools. These small cylinders came to be known as wampum, from the Narragansett word *wompi* or *wompan* referring to white shell beads. A string of white beads was known as *wampumpeague* or *wampumpeake*. Dark beads, in shades of light purple to almost black, were called *sacki* or *suki* beads. Purple beads were made from the inner shell margin of quahogs. Although many species of quahog inhabit the eastern U.S. coast, only the northern quahog, *Mercenaria mercenaria,* which is found from Maine to New Jersey, produces purple pigment embedded in its inner shell. The biological function of the purple pigment within the quahog shell has remained a mystery. Often, coloration in organisms provides visual cues to other organisms; however, the purple rim on the inside of the quahog shell is hidden from view. The purple coloration may result from a metabolic waste product, often absent in young quahogs. Current scientific thinking suggests that the pigment may have a function in strengthening the shell along the margin that is particularly vulnerable to chipping during attacks by predators such as whelks. Another hypothesis proposes that the purple metabolic waste deposited in the interior of the shell may be toxic to predators and pests that have the ability to bore through or etch indentations or pits on the shell (Kraeuter and Castagna 2001) but this idea does not explain why the pattern of pigment deposition is only on the outer margin of the inner shell surface.

The northern quahog received its official name *Mercenaria* because of the importance of the shell to Europeans as a form of currency. European coins were scarce, so wampum was adopted as a medium for trade, leading to the designation of beads as "Indian money." White beads were made from a variety of molluscs but most commonly from Atlantic coast whelks in the genus *Busycon*. The knobbed whelk, *Busycon carica,* and the channeled whelk, *Busycon canaliculatum,* have shells that contain an inner cylindrical core or whorl, a spiral columella, that lends itself to the making of beads. The relatively large size of whelks allowed several beads to be fashioned from each shell.

The production of cylindrical beads from bits of quahog shell was a more labor-intensive process. Only one or two beads could be produced from the anterior margin of each quahog shell. Because of the difficulty of producing the purple beads, they were not very common until colonial times, when more sophisticated tools, such as metal drills, became available.

When polished to a high luster, wampum beads, especially the dark ones, have a gem-like quality. Each purple bead takes on its own characteristics due to range of color (from the lightest lavender to almost black), depth of color, uniformity of color, and presence of striations. Aside from their intrinsic beauty, wampum beads were valued by native people because of the time, skill, patience, and labor involved in their production.

Shells were collected in the warmer months by the Narragansetts, Montauks, Mohegans, Shinnecocks, Pequots, Niantics, Quinnipiacs, Wampanoags, and other tribes from Cape Cod southward along the Rhode Island, Connecticut, and Long Island coasts to the Hudson River Valley. Only coastal tribes had access to whelks and quahogs. The shells and shell fragments were the raw material used for winter artistry. The shell bits, shaped into rough cylinders or cubes, were kept immobile by splinting them into a vice made from a split branch. From one side, a hole was made halfway into the shell with a stone or reed drill. Water was applied to keep the shell from heating up and cracking. The same procedure was used to drill a hole from the opposite side of the piece. Evidence for the technique of drilling from both sides come from X-rays of wampum taken by Dr. Elsie Fox in New York City that show that the holes through the long axis of the cylinders do not exactly line up near the center (Orchard 1975). Hand techniques using stone tools were employed to smooth the pieces into polished cylinders. Older, pre-colonial wampum is chunkier than that made in the 1600s. By the time Europeans arrived, the process had been refined to produce beads that were about 6 mm (1/4 inch) long and had a diameter of about 3 mm (1/8 inch). Because most of the quahog shells we see today are from very young clams, it is difficult to imagine a purple shell margin that is thick enough to produce a bead. However, quahog shells thicken with age. When quahogs were abundant, many of the older, thicker-shelled clams were readily available.

The beads were originally simply strung on plant fibers or animal sinew as single strands. The order and ratio of the white and purple wampum

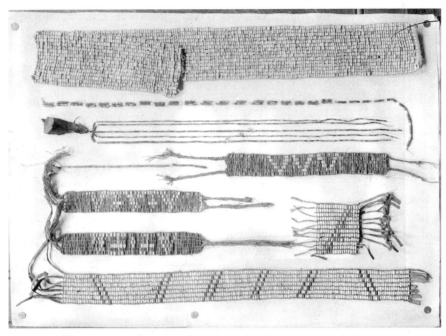

FIG. 4.5 Wampum items. *Photo: National Anthropological Archives; Smithsonian Institution.*

conveyed specific messages (Orchard 1975) and different tribes wore strings with specific color combinations to represent their identity during council meetings with other tribes. By the 1600s, after European tools were introduced and wampum production was in high gear, more elaborate wampum ornamentation was being produced. Initially, the tribal men appear to be primarily responsible for the manufacture of the beads. It is not clear whether men, women, or both were involved in stringing the beads or in producing the sometimes elaborate weavings. Purple and white beads were made into collars, bracelets, necklaces, belts, earrings, aprons, and other items (fig. 4.5). Although many of the items were meant to be worn, their purpose was not simply decorative.

Wampum had deep significance on many levels and was an important vehicle for communication. White wampum, symbolizing light and brightness, was featured in items that conveyed joy, friendship, and family bonds such as marriage, while dark or black wampum was the predominant background color in usage for more formal, serious, and somber occasions such as mourning or declarations of war. Dual-color designs or pictograms represented important occasions or events such as election or deposition of a tribal chief, mourning, religious ceremonies, tributes, ransom for captives, prizes for contests, declarations of war, signs of peace, tribal alliances, treaties, political relationships, and important tribal ceremonies. From an individual perspective, wampum was a sign of social status or badge of office.

It is important to note that wampum never was conceived of or used as currency by Native Americans. Although it may have been exchanged by tribes and used in trade between tribes, the significance of wampum was in its symbolism, not its monetary worth. As Martien describes it, wampum was "message and ornament, record and symbol, article of exchange" (Martien 1996, 24). For the natives, wampum was "a medium of ritual language" (78). This notion of sacred, symbolic wampum changed with the coming of European settlers in the 1600s. In 1609, when Henry Hudson arrived in the "New World," he received a string of wampum as a welcome gift from a tribe that lived along the river that was later named after him. He was not aware of wampum's potential value. In 1622, a Dutch fur trader from the Dutch West Indian Company, Jacob Eelkes (sometimes spelled Jacques Elekens) kidnapped a sachem named Tatobem from the Long Island Pequots. After threatening to behead the sachem, he received long strings of wampum, perhaps amounting to 35,000 beads, as ransom (Scozzari 1995). Tragically, the sachem was killed despite the ransom. The astute Eelkes soon discovered that wampum, needed to produce belts and other sacred ornaments, was a rare commodity away from coastal locations and that he could use it in trade with upriver natives for furs.

Native Americans initially had little need for many of the European goods proffered for trade but over time came to depend upon relatively inexpensive European trade goods such as blankets, fabric, metal tools, guns, and gun powder. With the realization that the native tribes valued wampum, Dutch traders sought to acquire as much wampum as possible from coastal tribes to use in trading with inland tribes for more valuable items such as furs. Beaver pelts, which were much sought after to make hats, could be sold for top dollar in Europe. The monetary value of wampum thus was tied intimately to the fur trade, hence the birth of wampum as a significant type of New World currency. For European colonists, it became "ready cash" (Martien 1996, 78).

The Dutch profited by obtaining furs, which were sold in Europe, from inland tribes in exchange for wampum, which they were able to obtain with relatively little expense from coastal tribes. The Dutch thus initiated a trade triangle: They provided European goods to the Pequots and to coastal tribes of the Connecticut River and Long Island Sound, and used the wampum in exchange for furs obtained from inland tribes at Fort Orange, their station on the Hudson River. Learning from the experience of the Dutch, the British also became involved in the wampum trade. The British wampum trade triangle included the fur-supplying Nipmucs in Massachusetts and central Connecticut and the bead-producing Niantics who loved in southern Connecticut and Rhode Island (Scozzari 1995).

By 1627, the British colony at Plymouth had established a trading post in Bourne, Massachusetts, at Apucxet. Apucxet was a convenient location because goods could be shipped down the Scusset River from Plymouth, carried over a short distance, and then transferred to small boats on the Manomet River. From the Manomet, the goods would be brought to the trading post at Apucxet, where they could be loaded onto Dutch ships in the deeper waters of Buzzards Bay and taken to New Amsterdam. Today, it is difficult to visualize the historical importance of this trading post. Although the building maintains its original appearance, modern life is close at hand. The Cape Cod Canal, built in 1914 and widened in 1940, runs immediately outside the trading post and can now quickly and easily handle any ship that makes the trip from New York to Plymouth.

Dutch trader Isaac de Rasiere (sometimes spelled DeRosieres or De-Razier), secretary of New Netherland, was sent to Plymouth by Peter Minuit as a good-will ambassador whose mission was to investigate the possibility of trade. De Rasiere brought clothes, linen, and white sugar to trade, along with some wampum, which the Dutch called *sewan*. (Other spellings of the Dutch name for wampum were *seawant, zeewant, zeewand, seewant, seewan,* and *seawan*.) He had heard that wampum was in great demand among the tribes along the Hudson and conjectured that it also would be valuable in New England (Scozzari 1995). To promote and expedite the three-way trading between the Dutch, British, and natives, de Rasiere proposed that wampum be used as a medium of exchange. Wampum was traded by the fathom, a rope of 6 feet containing about 360 beads, with purple beads more valuable than white ones (Jordan, 1998).

Wampum, as a critical link in the fur trade, changed the lifestyle and economics of the region. Fur-bearing animals were hunted more intensively and some tribes became more specialized in the making of wampum. Recognized by the Durch as a major center for the production of wampum, Long Island was called Seawanhackey, Sewhounhockey, or Sewounhockey; place of seawan or land of shells (Jordan 1998). In exchange for wampum and furs, native people became increasingly dependent on goods from Europe. By the 1620s, wampum was adopted officially as currency in New Amsterdam. In Massachusetts, wampum became legal tender in 1637, with a value of six beads to a penny. Wampum could be used to purchase land, pay taxes, and even pay tuition to Harvard College (Scozzari 1995)!

Wampum was difficult to produce; perhaps 36 to 48 beads could be produced per person per day (MacKenzie et al. 2002). It became so valuable that imitation beads, made of diverse materials such as glass, stone, bone, horn, and wood began to appear in circulation. The increased volume of trade led to an increased demand for wampum and hence increased production. Industrious colonists "took the opportunity to literally and figuratively, make some money, by producing 'counterfeit' wampum" (Jordan 1998). European tools, such as metal awls known as "muxes," eventually replaced traditional stone tools and allowed holes to be drilled straight through the beads from one end to the other. Awls became very valuable tools. A deed, signed in 1650 when the colonists acquired Huntington, Long Island, from the natives, mentioned that 30 muxes and 30 needles were used in trade. A similar deed for East Hampton, Long Island, in 1648 specified 100 muxes, and the deed for Mastic Neck, Brookhaven, Long Island, dated 1657, cites 40 needles and 40 muxes (Beauchamp 1901; Orchard 1975).

The decline in the fur trade and the difficulty of regulating the amount of wampum in circulation as well as its quality led to the eventual decline of wampum as the major unit of currency in the 1660s. Although coin replaced wampum as currency, wampum was still produced and used in trade throughout the nineteenth century. The Campbell family started mass production in 1812. They depended on the commercial clam industry in Jamaica Bay and such outlets as the Fulton Fish Market to supply the raw material to produce wampum at their factory in Passouk, New Jersey (now Park Ridge), for close to a century. The factory produced about 1,200 beads from a bushel of quahog (about 300 clams) (MacKenzie et al. 2002). Today, wampum is still being manufactured and sold for the production of jewelry and other ornaments (plate 13). Descendants of Native American tribes are reviving the manufacture of wampum beads, jewelry, and belts. The Wampum Theater at the Mashantucket Pequot Museum and Research Center at Foxwoods in Connecticut features a film segment that shows how wampum beads are made with modern tools. Despite the availability of power tools, the process is labor intensive and requires a great deal of skill.

FIG. 4.6 Tribal interpreter of the wampum: Engraving of painting by John Verelst of Tee Yee Neen Ho Ga Row (He Who Holds the Door Open), baptized Hendrick, and called Emperor of the Six Nations, holding wampum belt (1710). *Photo: National Anthropological Archives; Smithsonian Institution*

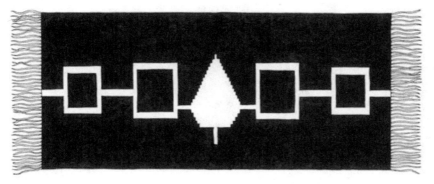

FIG. 4.7 Replica of Hiawatha belt. *Photo: Courtesy of John Fadden*

Wampum Belts

The availability of two contrasting colors, white and purple, allowed for the design of the beautiful, symbolic and pictorial designs found on wampum ornaments. A single belt might require thousands of tiny wampum. Starting with rows or strings of beads, native people produced wampum belts by weaving the rows together. Some of the more elaborate and detailed belts were produced by members of the Iroquois Confederacy after European settlement. The wampum belt was a sacred object embedded with an important message. It usually told a story and served as a memory aid in the passing of oral traditions. Often, one tribal member was the wampum keeper. He was entrusted with the belts and served as tribal interpreter (fig. 4.6).

One of the most famous and culturally important belt is the Hiawatha (Ayanwatha) belt (fig. 4.7). The wide, 38-row belt contains a dark background of 5,682 wampum beads; 892 white beads define the diagram of a central tree and the outline of two squares on either side of the central motif. All the squares and trees are attached to one another with white beads. The belt commemorates the alliance or Great Peace between five tribes of the original Iroquois Nation (which later became six nations after the Tuscaroras joined the Confederacy); the central tree represents the Onondaga, the keepers of the wampum. With the tree as reference point, the squares are found in approximate geographic orientation and represent the Mohawk (Eastern Door), Oneida, Cayuga, and Seneca (Western Door) tribes. Although many descriptions of the belt refer to the central motif as a pine tree, perhaps a location where tribal meetings were held, Martien's research and the interpretation of current Native Americans indicate that the belt

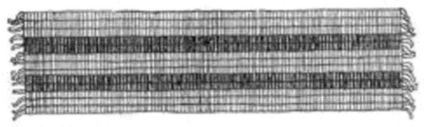

FIG. 4.8 Replica of Two Row Wampum belt. *Photo: Courtesy of John Fadden*

also can be inverted 180 degrees from the position in which it usually is depicted (Martien 1996; Tehanetorens 1999). The central motif then becomes a heart. This heart of the Five Nations is where the Council fire burns and where the Great Peace of the Five Nations is centered.

The Two Row Wampum Treaty belt commemorated a pact between the natives and the colonists (fig. 4.8). Two parallel rows of purple wampum on a white background symbolize the path of the natives, their customs and laws, and the separate path of the white man. "We shall each travel the river together, side by side, but in our own boat. Neither of us will make compulsory laws or interfere in the internal affairs of the other. Neither of us will try to steer the other's vessel" (Tehanetorens 1999, 74).

The Penn Treaty wampum belt was delivered to William Penn in 1682 by Lenni-Lenape (Delaware) chiefs (fig. 4.9). The belt depicts Indian and white man with hands joined in friendship. The To-ta-da-ho belt is a very wide belt, 27 inches long and 14 inches (45 rows of wampum) wide (fig. 4.10). Fourteen small, white diamonds are displayed against a series of overlapping purple triangles. The To-ta-da-ho belt sometimes is referred to as the Presidentia belt because it was the first belt held up by the chief at important council meetings of the Six Nations. The chain of diamonds is thought to represent a chain of friendship, or covenant. The

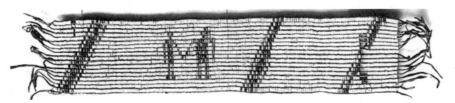

FIG. 4.9 The belt of wampum delivered by the Indians to William Penn at the "Great Treaty" under the elm tree at Shackamaxon, in 1682. *Library of Congress, Prints and Photographs Division (LC-USZ62-86486)*

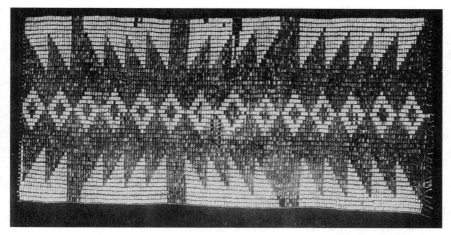

FIG. 4.10 To-ta-da-ho belt—diamonds in center said to be a covenant chain signifying alliance of towns. *Library of Congress, Prints and Photographs Division (LC-DIG-ggbain-00160)*

belt has been shortened, because it originally consisted of sixteen diamonds (Clark 1931).

Many of the original wampum belts found their way to museums or collectors of Native American artifacts. Unfortunately, some have disappeared or have been destroyed. The University of the State of New York became "Keeper of the Wampum" in 1898 under a rather curious arrangement. Prior to that date, the Onondaga Nation was "the keeper of the wampum and the wampum record of the Iroquois Confederacy," but under somewhat mysterious conditions involving the exchange of money, the Onondaga ceded the honor to the state (Clark 1931). At one time, the Museum of the State of New York in Albany had 25 belts, some of which were among the most important and valuable belts in existence (Clark 1931). Wampum belts of historical value are no longer on display or archived in most museum collections because technically, they are no longer property of the museums. As a result of the National Museum of the American Indian Repatriation Act of 1989, the Smithsonian and other museums have been actively involved in the return of certain items in their collections to lineal descendants and culturally affiliated Indian tribes, clans, villages, and organizations. Beginning in the 1990s, the New York State and other museums have returned sacred items such as wampum belts to tribal leaders. Today, the Onondaga Nation is again Wampum Keeper for the Haudenosaunee, the six nations of the Iroquois Confederacy.

With the resurgence of interest in Native American culture by descendants of the original inhabitants, many of the belts have been reproduced faithfully. Under a project supported by the America the Beautiful Fund of New York, New York State Council of the Arts, and the National Endowment for the Arts, students at the Indian Way School at Akwesasne Mohawk Nation have reproduced the belts, bead by bead, and have recorded their descriptions and meanings (Tehanetorens 1999). Some of the reproduction belts are on display at the Six Nations Indian Museum in Onchiota, New York.

Although the bivalve shell is its most notable feature and the use of the shell has important historical significance, today we are more interested in the edible parts of the bivalve that contribute to their economic value as well as other internal structures responsible for their growth, survival, and role in coastal ecosystems. We will consider bivalve anatomy in the next chapter.

PLATE 1 Spat on cultch.
Photo: Barbara Brennessel

PLATE 2 Eastern oyster, *Crassostrea virginica.*
Photo: Scott W. Shumway

PLATE 3 Hard clam or quahog, *Mercenaria mercenaria*.
Photo: Barbara Brennessel

PLATE 4 Softshell clams, *Mya arenaria*.
Photo: Scott W. Shumway

PLATE 5 Sea scallop, *Plactopecten magellanicus*.
Photo: Nina Picariello

PLATE 6 Bay scallop, *Argopecten irradiens*.
Photo: Nina Picariello

PLATE 7 Blue mussel, *Mytilus edulis.*
Photo: Scott W. Shumway

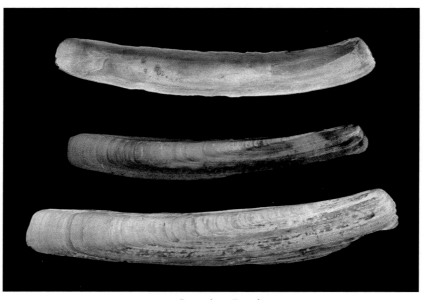

PLATE 8 Razor clam, *Ensis directus.*
Photo: Scott W. Shumway

PLATE 9 Surf clam, *Spisula solidissima. Left* shell shows inside with adductor muscle scars and pallial line. *Right* shell exhibits residual periostracum layer.

Photo: Nina Picariello

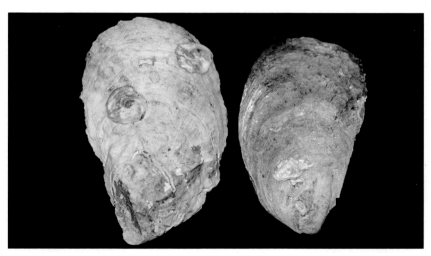

PLATE 10 Oyster shells.

Photo: Nina Picariello

PLATE 11 *M. mercenaria notata* with zig-zag shell patterns.
Photo: Barbara Brennessel

PLATE 12 Inside the quahog shell, showing purple margins used for production
of wampum and some modern wampum items.
Photo: Nina Picariello

PLATE 13 Necklaces made by Chatham shellfisherman Martha Nunuz, who collects large quahog shells from her 32-foot dragger in Nantucket Sound.

Photo: Courtesy of Martha Numez

PLATE 14
Bay scallop's blue eyes.
Photo: Adriana Picariello

PLATE 15 Cultures of unicellular algae, used to feed broodstock and larvae in the hatchery.
Photo: Barbara Brennessel

PLATE 16 Chinese hat covered with oyster spat.
Photo: Paul Bonanno

PLATE 17 *Codium.*
Photo: Barbara Brennessel

Chapter 5

Core Values
A Primer of Bivalve Internal Anatomy

After a bivalve mollusc is opened, it is sometimes difficult to distinguish the anatomical structures that are characteristic of this group of animals. Careful observations have led to the description of internal structures and elucidation of their functions. The internal structures of the oyster (fig. 5.1), quahog (fig. 5.2), and blue mussel (fig. 5.3) can be compared and used as reference in considering the important physiological roles of these tissues and organs.

Muscles

The valves of molluscs are joined at the hinge area, near the umbo. Indentations where the valves join are known as teeth, which interlock to form a tight joint, similar to a tongue-in-groove effect. Ligaments in the hinge region allow the shell to spring open when the valves are not kept closed by the major muscles, the adductors. The sites of adductor muscle attachment to the shell are seen as scars on the interior surface of the shell (see plate 9). Clams have two large adductor muscles, one on the right side and one on the left. Mussels also have two adductor muscles; the one near the hinge is much smaller than the one near the margin of the shell. Scallops and oysters only have one, large, centrally located adductor muscle. A portion of the adductor, described as "quick" muscle, causes the shell to close when it contracts; another portion, the "catch" muscle, keeps the valves closed or partially closed. The muscle of the scallop is very large because it is pumped up by frequent swimming activity and is the only part of the animal that is eaten and sold commercially in the United States. In Asian countries, several scallop preparations involve the whole body of the animal.

The adductor muscles have the important function of keeping the valves closed, but they have an additional role in the scallop: they are responsible for swimming movement. Scallops clap their valves with a rapid open and

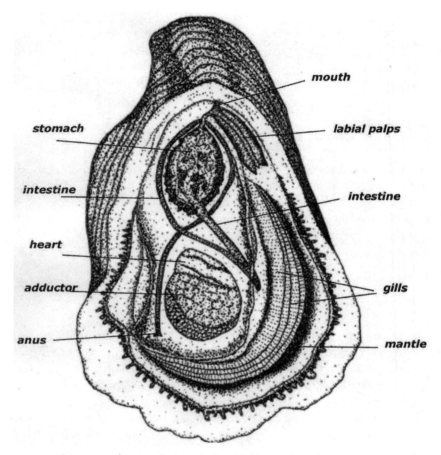

FIG. 5.1 Oyster internal anatomy. *After Pechenik, modified from several sources. Drawing: Marisa Picariello*

close motion that projects the animal backward. Scallops also can eject water from the margins of their shells to move around and flip themselves right-side up if they have been turned upside down. The well-developed adductor muscle has a good supply of glycogen, a polysaccharide that functions as a stored form of energy that is metabolized when scallops require this reserve for production of gametes. When we eat scallops, muscle glycogen is broken down to the simple sugar glucose by the enzyme amylase in our saliva, thus accounting for the sweet taste and culinary popularity of this bivalve.

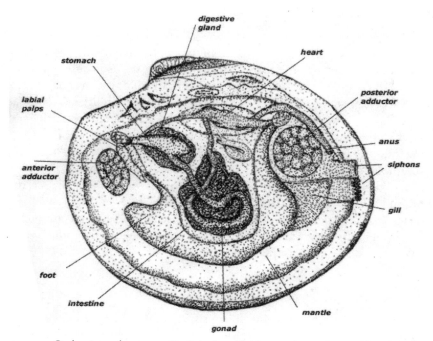

FIG. 5.2 Quahog internal anatomy. *After Pechenik, modified from several sources. Drawing: Marisa Picariello*

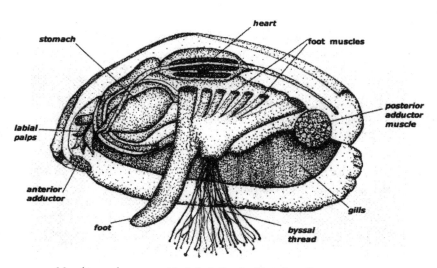

FIG. 5.3 Mussel internal anatomy. *After Pechenik. Drawing: Marisa Picariello*

The Mantle

Bivalve molluscs have a thin, translucent tissue, the mantle, that surrounds their internal organs. The mantle of the blue mussel also contains its gonads. The mantle is not an inert bag that simply holds the internal organs in place. It has many vital functions, including sensory abilities. Its musculature allows it to sweep the inner surface of the shell to remove particles of sand and debris. The inner mantle fold helps control the flow of water—the source of food and oxygen—in the mantle cavity. The middle fold is responsible for sensory functions and contains structures such as the blue eyes of the scallop (plate 14). The most important role of the mantle, is the production of shell, mediated by the outer fold of the mantle which secretes shell material that it produces by concentrating calcium compounds dissolved in the water. The mantle is also responsible for the production of valuable molluscan gems: pearls.

Pearls

Pearls are predominantly mineral, consisting of the shell material calcium carbonate in two main forms, calcite and aragonite. However, the components of the two forms differ in spatial arrangement. Pearl-producing oysters have a unique nacreous layer lining their shell, consisting chiefly of aragonite material (nacre or mother-of-pearl) and also the protein/polysaccharide complex known as conchiolin. The formation of a pearl is an extension of shell formation. The shell-forming-tissue of the mollusc coats a foreign object that cannot be dislodged. Most people assume that pearls are formed in response to grains of sand that irritate the mollusc. In natural pearls, the foreign object, referred to as a nucleus, may be a parasite, a piece of tissue, a tiny marine organism, or a small piece of organic matter.

Pearls, the only gems produced by living animals, are made by many species of mollusc. However, luminous, gem-quality pearls are not produced by molluscs that have important value for their culinary properties. Because of the value of pearls throughout history, most of the mollusc species that produce gem pearls have been overharvested. In order to keep up with demand, techniques were developed to induce certain species of freshwater mussels and oysters to produce pearls. In cultured pearls, a small bead, made from the shell of another mollusc, serves as the nucleus. Originally considered to be second-rate gems, cultured pearls have now gained wide acceptance.

The common molluscs that are fished or farmed in the Northeast, such as hard clams, surf clams, ocean quahogs, oysters, softshell clams, bay and sea scallops, and blue mussels, have no value in the commercial pearl industry. The lumpy pearls that are produced occasionally by food bivalves actually are considered to be a nuisance, potentially the cause of a broken tooth! Occasionally there is an exception, as exemplified by an event that occurred on a stormy December day in 2005 when a Rhode Island woman had a craving for clams. She braved blizzard conditions and made her way to the local seafood store in Newport, where she purchased several dozen quahogs. As her husband shucked the clams, he noticed a strange-looking lump inside one of them. Thinking he had a sick clam, he was about to discard it when a second look revealed the treasure within. It was a lustrous purple pearl about the size of a large pea. The actual worth of the quahog gem is not known because quahog pearls are relatively rare and not commonly sold. Some collectors have speculated that the value of this purple pearl may be tied into an item of jewelry that was purchased in a Rhode Island antique store: a broach fitted with two quahog pearls that was purchased for $14.00 but now may be worth close to a million dollars.

Siphons

Infaunal bivalves—those that live in the sediments—need access to water and the currents that supply food and oxygen. Specialized parts of their mantle fold into tubes known as siphons that can extend above the surface of the sediment and allow these bivalves to feed and obtain oxygen while remaining protected. The paired siphons sometimes are referred to as "necks," hence the names "little necks" for small quahogs and "long necks" for softshell clams (fig. 5.4). Razor clams have relatively long siphons that allow this species to burrow deeper into the sediments than either quahogs or softshell clams. One siphon is for input (the incurrent siphon) while the other is used for output of water that already has been filtered and contains waste (the excurrent siphon). The siphons of the quahog can be retracted completely into the shell, while those of the softshell clams always remain at least partially exposed, while covered with a thin, brown protective coating of periostracum. Epifaunal bivalves, which remain above the sediments—oysters, scallops, and mussels—do not have siphons but nonetheless have an incurrent and excurrent pattern of water flow through the body cavity. Water intake occurs as the shells open slightly or "gape."

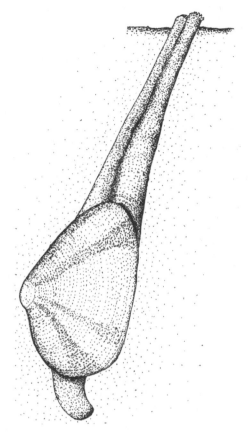

FIG. 5.4 Long neck of softshell clam. *After Pechenik.*
Drawing: Marisa Picariello

Gills

Gills are dual-function organs. They are key components of the filter-feeding mechanism of bivalves and also are employed in the process of gas exchange. Thus they have a very complex structure and function. Each bivalve mollusc has a pair of gills that are folded to maximize surface area into a structure that has the shape of the letter "W." Each half of the "W" is referred to as a demibranch. The gills are covered with thousands of tiny hair-like structures known as cilia whose movement creates a current that directs the flow of water though the gill and removes particles from suspension. As the particles are channeled through the conduits of the gill to the food grooves, they become coated in a slimy mucus material produced by cells in the gill. The mucus prevents loss of the food particles from the food channels. The particles ultimately are collected in the food groove and then conveyed to the labial palps, where they are directed to the mouth and digestive system.

Size is not the only factor in food selection. Bivalves can discriminate among types of food and display a high degree of selectivity. The composition of the food appears to be the key factor in the ability of the bivalve to select particles for ingestion, but very little is known about the exact characteristics a bivalve seeks in its food supply. Food particles that are rejected are formed into aggregates called pseudofeces and expelled. Although they are waste products of a sort, the aggregates have not traveled through the digestive system and are not technically considered to be fecal matter.

Bivalves are well known for their phenomenal filtration capacities. The Chesapeake Bay Foundation reports that a single oyster can filter 60 gallons (227 liters) of water each day, and that in prior times, the entire water volume of Chesapeake Bay, amounting to 19 trillion gallons of water, could be filtered in a week by the resident oysters (Chesapeake Bay Foundation n.d.). It has been estimated that with today's oyster populations in the bay, it would take over a year to achieve the same level of filtration. The exact filtration capacity of a single bivalve depends on a number of factors, including temperature, size of the animal, and the manner and conditions under which filtration is measured. In a comparative study under controlled laboratory conditions, an average-sized oyster was able to filter approximately 2 liters (2.1 quarts) per hour. The average bay scallop filtered about 3.4 liters (3.6 quarts) per hour, while the quahog filtered 1.24 liters (1.3 quarts). Biologists speculated that the quahog might filter twice as much water if the rate was measured in situ, under sediments in natural conditions (Riisgard 1998).

The Digestive System

The main structures of the bivalve digestive system are the labial palps, mouth, esophagus, stomach intestine, and rectum. The labial palps (sometimes referred to as lips) at the entrance to the mouth are responsible for additional food tasting and sorting and are capable of discriminating among food particles and directing the preferred food particles into the mouth. The rest of the digestive system has functions that are fairly similar to those of other animals. A noticeable exception is a structure known as the style. This structure is associated with the stomach and consists of a crystalline style that is contained within a style sac. The crystalline style is not at all crystalline. The name is a holdover from early observations of this clear, glass-like structure and misconceptions about its actual make up. The style is composed essentially of protein and polysaccharide and is rotated by hair-like cilia in the stomach. The rotational mechanism appears to be important in the breakdown of dietary algae and silica-coated diatoms. The rotation of the style grinds the food particles and mixes them with an array of digestive enzymes (Rice 1992). A fascinating feature of this structure is that its size varies with environmental conditions such as tides and current flow. It is a dynamic structure that is constantly being dissolved and reconstituted (Newell and Langdon 1996; Beninger and LePennec 2006).

The Nervous System

When we use the expression "happy as a clam," are we assuming that clams experience emotions? We also might assume that emotion is mediated by specific areas of the brain and that the brain would be located in the head. Some molluscs, such as squid, have a body plan with a well-defined head and tail, but during evolution, the bivalves became ancephalic, that is, they lost their heads! Bivalves have a simple nervous system that consists of three pairs of ganglia (dense, organized bundles of nerve cells)—cerebral, pedal, and visceral—connected to peripheral nerves. This pared-down nervous system mediates all the sensory functions of the animal and coordinates the movement of the valves, but it is unlikely that it is responsible for emotion of any kind. The expression "happy as clam" makes no sense unless we explore its origin, which was the complete phrase, "happy as a clam at high tide," which may allude to the fact that clams usually are harvested at low tide and are relatively safe from the rake at high tide. Perhaps the expression also reflects the fact that clams only feed when they are under water, so they would be most "content" at high tide. The expression apparently was well known by the mid-1800s, as evidenced by its inclusion in the John G. Saxe poem "Sonnet to a Clam":

> Inglorious friend! most confident I am
> Thy life is one of very little ease;

Albeit men mock thee with their similes
And prate of being "happy as a clam!"
What though thy shell protects thy fragile head
From the sharp bailiffs of the briny sea?
Thy valves are, sure, no safety-valves to thee,
While rakes are free to desecrate thy bed,
And bear thee off—as foemen take their spoil—
Far from thy friends and family to roam;
Forced, like a Hessian, from thy native home,
To meet destruction in a foreign broil!
Though thou art tender yet thy humble bard
Declares, O clam! thy case is shocking hard!

Of particular interest are the blue eyes of the scallop, which sit atop short stalks on the mantle, near the edge of the shell (plate 14). The eyes are simplified compared to those of higher organisms such as vertebrates, but they consist of all the components required for visual perception—cornea, lens, and retina—and they communicate with peripheral nerves. The eyes allow scallops to respond to shadow and movement, stimuli that elicit shell closure and swimming responses, behaviors that come in handy in avoiding predators such as starfish.

The Circulatory and Excretory Systems

Bivalves have a three-chambered heart and a blood supply known as hemolymph. The hemolymph does not contain special proteins such as hemoglobin to carry oxygen. Instead, the oxygen is dissolved directly in the fluid as it circulates through the system. The circulation is partially mediated by vessels, but it is not a closed system. In certain regions of the body, the hemolymph fills areas called sinuses and delivers oxygen and nutrients as it flows around tissues before it is collected again into other vessels (Eble and Scro 1996; Beninger and LePennec 2006; Rice 1992).

Within the hemolymph are various types of cells such as hemocytes that have multiple functions. They are capable of phagocytosis (ingestion) of potential parasites and thus are involved in the bivalve's immune defense system. They also are capable of transporting nutrients to other tissues, as well as removing waste and transferring it to the kidneys. In addition, hemocytes are involved with repair processes. For example, they can deliver shell materials to sites of breaks and they can clump at the site of a wound to wall off the area from infection.

Bivalves produce urine as a waste product in their kidney. Several nitrogen-containing waste compounds are produced, the most predominant being ammonia.

The Foot and Byssus

During the early developmental period, all bivalves have a foot, a muscular organ that is used for seeking suitable substrate for settlement. The foot is lost in oysters and scallops, but it remains an important structure in adult clams and mussels. In clams, it is used for burrowing and thus allows clams to move up and down or back and forth in the sediments. The movement can be remarkably and unexpectedly rapid, as anyone who digs for steamers or razor clams can attest. The length of the siphon determines how deep a bivalve can burrow and still continue to feed.

The foot also contains the glands that are responsible for the formation of byssal threads, which allow molluscs in their early developmental stages to attach to substrates. The production of these attachment threads, collectively known as a "byssus," is an important function that is lost in quahogs and oysters but retained in scallops and blue mussels, where it is critical for the survival of these organisms.

Imagine a thread that is made from the same material as human tendons, strong and stiff at one end and elastic and pliable at the other yet pound for pound as tough as Kevlar, the material that is fashioned into bullet-proof vests for law-enforcement agencies and the military (Coyne, Qin; and Waite 1997; Brazee and Carrington 2006). This structure is a byssal thread that makes up the byssus of bivalve molluscs, necessary for attachment of the organism to substrates. The byssus, from the Latin term for fine linen, is formed during early development and provides an important anchor but need not be permanent; it can be cut and reformed, allowing bivalves to move to another, more preferable location.

The blue mussel retains its ability to produce byssal threads throughout its life. Blue mussels depend on the threads to secure them to substrates or structures so that they can withstand the displacement powers of waves, currents, and tides. This strategy of attaching or anchoring in a favorable habitat is similar to that used by other sessile molluscs such as oysters, barnacles, and chitons. These marine organisms all have very short or absent siphons, so the attachment to substrates allows them to remain above sediments and near their food and oxygen supply. Although the strategy is similar, the mechanism for attachment of mussels is markedly different. Oysters, barnacles, and chitons produce a cement-like material that glues them firmly to the substrate. In contrast, the mussel byssus is very flexible and functions more like a rope that provides attachment but allows some changes in orientation in the water column, so that the mussel can take advantage of the direction of its incoming food supply.

The byssal threads are composed of collagen, the same type of protein that is contained in connective tissues such as bones, tendons, and ligaments of many animals. The threads are stronger than tendons yet more elastic and flexible. The material, produced by the byssal gland at the end of the foot, belongs to a familty of proteins known collectively as mussel foot proteins. The byssal gland extrudes the protein, which is then molded into fibers and coated with a tough cuticle for protection (Brazee and Carrington 2006). The entire process for the production of a byssal thread takes about two to five minutes. The thread itself has three main regions that make it strong yet resilient. One end has components similar to those that make up the silk proteins produced by spiders; the other, more elastic end has components similar to the rubber-like animal protein known as elastin, while the middle region is strong and inflexible (Lucas, Vaccaro, and Waite 2002; Brazee and Carrington 2006). The overall nature of the structure has been described as a stiff tether at the substrate attachment end and a shock absorber with 160 percent extensibility at the end extending from the mussel (Coyne, Qin, and Waite 1997). Many laboratories are studying the mechanism of formation, tensile strength, and mechanical properties of the byssal threads to gain an understanding of their force and flexibility and to explore possible commercial applications.

In addition to the tensile strength of the byssal threads, scientists have been very interested in the plaque proteins or "mussel glue." This adhesive is found at one end of the byssal thread and is responsible for the attachment to a substrate. Mussel adhesive plaque protein is a marine super glue that attaches to wet surfaces, cures or sets very rapidly, and stands up to salt water. Curiously, the gene that codes for the plaque proteins resembles genes in a family that code for the growth hormone known as Epidermal Growth Factor (EGF) (Inoue et al. 1995), a protein with an unrelated function. There is tremendous interest in mussel adhesive plaque proteins and their bonding properties because of possible applications in marine environments, medicine, and dentistry, fields in which water-resistant bonding materials are always important. Further benefits of an adhesive from a natural source such as mussels are the probability that it would be environmentally safe and the expectation that it will be biocompatible within the human body. It has been possible to extract mussel adhesive protein from byssal threads, but the yield has been so low that most commercial applications are not feasible. On a small scale, the adhesive is being extracted in laboratories and marketed as CellTak in milligram quantities as a tool to help researchers attach cultured cells to inert supports such as glass slides or plastic dishes. Some scientists are exploring the possibility of mass production of the protein using the tools of recombinant DNA technology, similar to the commercial method of production of insulin, used to treat diabetes. The finding that mussel foot proteins contain significant amounts of the unusual amino acid building block L-3,4-dihydroxyphenylanlanine (DOPA) has intrigued scientists. DOPA, which is used to treat the symptoms of Parkinson's Disease in humans, forms strong cross-links when oxidized and thus may be responsible for some of the strength and adhesive properties of byssal threads and plaque.

Gonads

Gonads are reproductive organs that become pronounced during specific reproductive periods of the year. To appreciate the structure and understand the function of these organs, it is important to consider them in the context of the reproductive cycle of each species, which we will discuss in chapter 6. As we understand the reproductive cycle, we can appreciate the strategies used for modern hatchery production of bivalves.

Chapter 6

Perpetuation

Reproduction and Early Development

Bivalve molluscs have interesting sex lives. Biologists generally observe that most Eastern oysters (*Crassostrea virginica*) and northern quahogs (*Mercenaria mercenaria*) begin their reproductive lives as males but usually change into females after the first reproductive cycle. This type of reproductive mode is referred to as dioecious (having male and female structures on the same organism), with alternate hermaphrodism or protandrous hermaphrodism (first male). There are always some exceptions to the rule but it is difficult to estimate how many oysters or clams may start out as females (protogynandry). The change from male to female requires an extreme makeover of the reproductive gland from a testis to an ovary. In general, the youngest mature specimens will be males, the older specimens will be females, and those of middle age may be a mixture of males and females. You can't discern the sex of a clam or oyster by its appearance, even after it is opened. It is quite difficult to tell whether a clam or oyster is serving its reproductive duty as a male or female except by a microscopic observation of the gonads to determine the type of gamete—egg or sperm—that is being produced.

Gametogenesis

Some analysis has been done of the value of this type of reproductive strategy for the mollusc. Bivalves draw upon their glycogen reserves, stored in adductor muscles, to supply the energy required for gametogenesis. During its early years, the mollusc is growing more rapidly than it will in later years. Most biologists believe that to elude predators, much of the material and energy being used by the organism must be diverted initially to growth and production of shell. During this time, it is easier energetically for the mollusc to produce sperm, gametes that are smaller and require less energy and material than eggs. Even as a young individual, "being male" allows the organism to join the reproductive ranks of the species while still paying attention to the need for continued growth. Once growth slows down, and the mollusc survives its critical early years, more reserves can be devoted to the production of eggs. In oysters, where it has been studied extensively, the sex-switching process is complex and has been linked to such environmental parameters as nutrient stress, parasite load, and the sex and proximity of nearby oysters. However, "once a female" does not mean "always a female." Some female oysters can revert back to being male during the period between spawnings when the gonad is undifferentiated.

Scallops are true hermaphrodites, having both male and female reproductive organs developing at the same time within each individual that is released in the same season. However, as a hedge against self-fertilization, male and female gametes usually are not released at the same time and male gametes usually are released first (Fay, Neves, and Pardue 1983). No matter whether the mollusc is male or female or both, each spawning event is characterized by the production of a great number of eggs and sperm. To ensure survival of the species, the number of gametes released by bivalves is phenomenal. Only a small percentage will be fertilized, and of those, an even smaller percentage will reach adulthood. It becomes a numbers game. Under laboratory spawning conditions, it is possible to count the number of gametes released by the average adult mollusc. For example, Loosanoff and Davis (1963) conducted carefully controlled experiments to quantitate the number of eggs released by a single bivalve over a period of two months. Each quahog produced between 8.0 and 39.5 million eggs, with an average of 24.6 million eggs per clam. The greatest single spawning event resulted in 24.3 million eggs. Under the same conditions, oysters produced 23.2 to 85.8 million eggs per animal, with an average of 54.1 million eggs during the two-month period. The numbers vary with other scientific reports, but in general, a quahog can produce 10 to 30 million eggs in a single year, while the female oyster has been reported to produce anywhere between 2 and 500 million. The large variation is due to experimental conditions, age of oysters, and the site from which oysters were obtained. Each bay scallop and softshell clam produces fewer eggs than a quahog or oyster, but they still number in the millions per year. The number of sperm is larger by several orders of magnitude and can number in the billions. In the north, bivalve molluscs will spawn once a year; several spawning cycles are typical in southern waters. Bivalves continue this tremendous output of gametes for their entire lifespan, although fecundity (the ability to produce many offspring) may be highest in midlife when the bivalve has achieved near-maximal size. The reproductive life of a bivalve mollusc is therefore just about as long as it natural lifespan, which can be about 40 years for quahogs and 10 to 12 years for oysters. The noncultivated ocean quahog may live and reproduce for over 100 years. Bay scallops have to get it right the first time. Since their lifespan averages 12 to 16 months, and maximum lifespan is just over two years, most bay scallops will reproduce only once, when they are approximately one year of age (Belding 1910; Fay, Neves, and Pardue 1983). Sea scallops don't produce gametes until their third year and have about a decade to perfect their attempts at reproduction (Hart and Chute 2004).

Spawning and Fertilization

Production of large numbers of gametes is necessary because the reproductive cells are released into the great abyss of water surrounding the bivalve and immediately are diluted, potentially swept by tides and currents, and may travel far from the mother or father mollusc. Egg and sperm are hoping upon hope to encounter the appropriate complementary gamete, initiate fertilization, and give rise to a baby mollusc. The release of egg and sperm in this manner minimizes the risk of self-fertilization and encourages cross-fertilization, a mode of reproduction that is better for the species in the long run; it promotes the mixing of genetic material that may produce individuals that have better survival traits. However, time is of the essence. Egg and sperm remain viable for only a few hours.

One oyster species follows a slightly different approach to reproduction. *Ostrea edulis,* a commercially valuable oyster in Europe and one that is cultivated extensively on the northwest coast of the United States, uses an internal fertilization mechanism. The female releases a smaller number of eggs than *Crassostrea virginica,* but they are retained on the gills. As the eggs are fertilized, they are brought back into the female with the current and undergo early development within special brood pouches. The larvae are released after they undergo early development, and thus have better chances of survival than eggs that are fertilized in the water column. *O. edulis* thus produces fewer eggs then *C. virginica* but nurtures them once they are fertilized. M. F. K. Fisher contrasts the two modes of reproduction as follows: "the Atlantic coast inhabitants spend their childhood and adolescence floating free and unprotected with the tides, conceived far from their mothers and their fathers too by milt let loose in the water near the eggs, while the Western oysters lie within special brood-chambers of the maternal shell, inseminated and secure, until they are some two weeks old. The Easterners seem more daring" (Fisher [1941] 1988, 3–4).

Some synchrony in the release of gametes also must occur in order for reproductive success. What good would it be for the male to release sperm one week and the female to release her eggs a week later or vice versa? To keep males and females on the same reproductive schedule, several factors come into play, with water temperature as one of the key signals. When water warms to a specific temperature, usually during the late spring, spawning will be induced. In the Chesapeake Bay, spawning begins in Eastern oysters when water temperature exceeds 20°C (68°F), is maximal when temperatures are 22.2°C (72°F), and stops when temperatures exceed 29.4°C (85°F). In the Northeast, oyster spawning will begin at 15.6°C (60°F).

The gametes of the northern quahog develop when water temperatures reach 10°C (50°F) but are not released until water temperature exceeds 20°C (68°F) (Rice 1992). There is still some debate whether spawning is triggered by a set temperature or by the rate of warming. Some researchers believe that gradual warming in spring induces gamete formation but rapid temperature change may trigger spawning. Either way, there is no doubt that gonadal development and spawning are tied to changes in water temperature. However, other events also occur when the water warms up, including increases in the phytoplankton that serve as food and nourishment. Thus, increase in amount or types of food is also an important element in the timing of reproductive events.

In clams and oysters, males spawn first, with gametes expelled through the excurrent siphon in clams and on the excurrent side of oysters. Billions of sperm form a white, wispy stream. In the Chesapeake, watermen often witnessed the waters turning milky near oyster reefs during spawning season (Galtsoff 1964). Eggs are released intermittently rather than as a single pulse.

In addition to temperature, some indications suggest that molluscs communicate with one another by the release of pheromones. Pheromones are "external hormones" produced by one organism but having an effect on another. Under artificial breeding conditions and also presumably in the wild, the release of gametes by one animal will cause a chain reaction. When a mollusc begins to spawn, its release of pheromones serves as an announcement to other molluscs that it is time to spawn and as an inducement for them to do likewise. These important observations on natural spawning cues are the basis for protocols employed in hatcheries that produce small clams, oysters, and scallops for commercial uses and restoration projects from mature broodstock. However, the role of internal hormones and the contribution of genetic factors to the bivalve reproductive cycle are poorly understood.

After spawning, bivalves are "spent." Their glycogen reserves have been depleted, their meat may be thin and watery, and thus they are not in good market condition. They generally pass through a resting stage before they can build up glycogen reserves, resume gonadal development or switch sexes, and produce another season's worth of gametes. Generally, the spent condition may prevail in late spring and early summer—months without "r." The notion that oysters should not be consumed during this time period may be related to the fact that this is the period when they are the least tasty. Other factors may have provided the impetus for the restriction on consumption in non-"r" months. In Europe, *Ostrea edulis* females brood their young during the late spring and early summer, so the restriction may have stemmed from a conservation strategy. In addition, late spring and early summer is the prime season for several types of red tide and proliferation of certain bacteria that can cause illness in oyster consumers.

Larval Development

Mostly by chance, a small percentage of the total component of eggs and sperm find each other in the water column, fertilization ensues, and development begins. All bivalve molluscs pass through similar early development stages (figs. 6.1 and 6.2). The earliest distinctive form of the mollusc, the larva, forms after the fertilized egg undergoes a series of cell divisions. It takes the fertilized eggs about 18 to 48 hours to become trochophore (from the Greek, meaning "wheel-bearer") larvae, tiny free-swimming creatures with thousands of tiny hairs known as cilia. For two to three weeks, the molluscs join the ranks of meroplankton, organisms that spend only their early developmental period as plankton. They move about in the water column but may not be dispersed uniformly. Several studies have attempted to determine whether the larvae of oysters are distributed passively or whether they actively move in response to stimuli such as light and salinity (Kennedy 1996). Overall, it appears that larvae are not completely passive. They tend to concentrate in the upper water layers but avoid the surface. Although they can swim, they are not strong swimmers and thus get swept about by tides and currents.

The digestive tract soon develops and the larvae become voracious. The trochophore then passes through several veliger stages. The term "veliger" refers to a wispy, sail-shaped organ, the velum, which functions in food capture, respiration, and motility. No matter the type of bivalve, the first

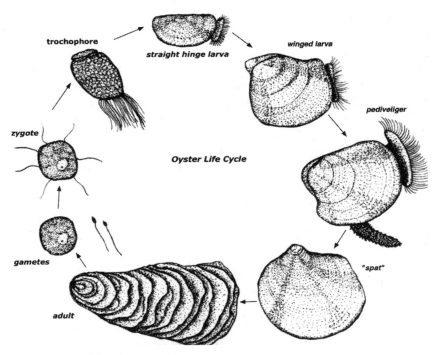

FIG. 6.1 Oyster life cycle. *Modified from several sources. Drawing: Marisa Picariello*

veliger stage is known as the straight-hinge or D-hinge stage. As the two shells develop, they are held together at one end with a hinge. The resulting larva resembles the capital letter D. Melbourne Carriker describes this stage in rhapsodic terms:

> A glance under low microscopic magnification of an actively swimming, straight-hinged veliger, its valves slightly parted, velum fully extended, rapidly beating cilia projecting widely beyond the edge of the glistening valves, brings into view an exquisite, vibrating icon. Clean, translucent, aragonitic valves shimmering mirror light as the mite spirals slowly in no particular direction, its velar retractor muscles twitching from time to time in response to a velar collision with some tiny planktonic particle. Not to have seen a living vibrant molluscan veliger is to have missed one of marine biology's larval treasures. (Carriker 2001)

If all goes well, the D-hinge veliger develops further as the mantle produces more shell material and the hinge becomes a miniature umbo (Ken-

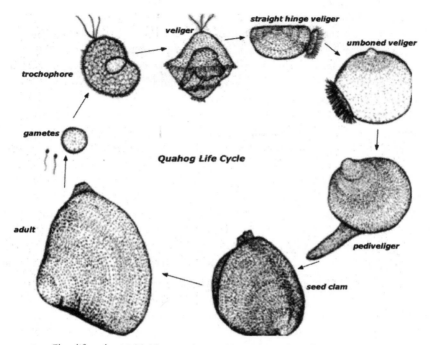

FIG. 6.2 Clam life cycle. *Modified from several sources. Drawing: Marisa Picariello*

nedy 1996). When larvae form enough shell material, the miniature shells initially protrude from their bodies, giving the appearance of wings and leading them to be called "winged larvae" during this developmental period. All bivalve mollusc larvae develop a foot and become pediveligers. The development of bivalves begins to diverge at this stage, and after a week or two, larval clams can be distinguished from larval oysters and larval scallops.

Early Clam Development

As a clam develops, the weight of its growing shell pulls it to the bottom, where it enters the benthic (bottom-dwelling) community and actively explores its new home, alternatively crawling and swimming, looking for a place where it can burrow and complete its development (see fig. 6.2; Carriker 2001). Clams prefer substrate that is sandy, a combination of sand and mud or a combination of mud and gravel, but they also may seek small rocks or other clams. In general, the nature of the setting cues for clams is still a mystery. If a setting clam finds suitable substrate, it will attach with a tiny byssal thread, woven by a special gland near the base of the foot. The byssus provides an anchor, giving the clam purchase on the bottom and protecting it from the turbulence caused by currents and tides. During this developmental period, the clam is called a plantigrade larva. It can attach, release, and reattach in another location. It may join the small percentage of larvae that eventually will be able to employ their feet to burrow into the substrate. As its siphon develops, it will lose its ability to produce a byssal thread.

Early Scallop Development

It takes about two weeks for a larval bay scallop to develop from the fertilized egg, reach the juvenile stage, and undergo settlement. Sea scallops may remain planktonic for over a month. Similar to larval clams, the late-stage scallop larva uses a byssus to secure its attachment to a suitable substrate and prevent it from sinking into bottom sediments where predators abound. Scallops differ from clams, however, in the developmental stage after the D-hinge: Scallops maintain a straight hinge, but develop "wings" (Cragg 2006.) Although a number of substrates can be utilized, the ideal settlement medium for bay scallops appears to be blades of common eelgrass (*Zostera marina*), but other algae, stones, and even oyster shells may serve as anchorage. Eelgrass is not an absolute requirement for settlement, but scientists believe it is a preferred attachment site. To remain out of reach of predators, settlement usually occurs on blades at locations about 12 to 15 cm (5 to 6 inches) from the bottom (Belding 1910). Sea scallops have been known to settle on shell fragments, rocks, pebbles, and other living organisms such as algae, larger scallops, and hydroids (Hart and Chute 2004). Juvenile scallops have been found to feed better when water currents are slower. Thus, in addition to preventing burial in the sediments and predation, eelgrass and other objects that are used for settlement may deflect currents and enhance feeding. The juvenile scallop is not attached irrevocably. It can detach, crawl, and reattach as it explores its environment during its dispersal stage and waits to develop its swimming abilities.

Early Oyster Development

The oyster larva will spend two to three weeks in the water column. After its D-hinge veliger stage, it will develop eyes and a foot and become a pedi-

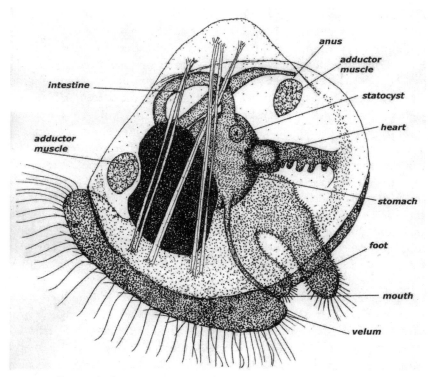

intestine

adductor muscle

anus

adductor muscle

statocyst

heart

stomach

foot

mouth

velum

FIG. 6.3 Oyster pediveliger. *Modified from several sources. Drawing: Marisa Picariello*

veliger (fig. 6.3). Its two pigmented eyes, known as statocysts, are not true eyes because they do not function in vision. They are believed to be sensory glands and may serve as gravity detectors. The foot of the larva helps it to explore the bottom to find the right place to settle down. This stage is sometimes called the eyed veliger stage, and is a sign that the swimming period is just about over and setting time is close at hand. An anthropogenic description of the oyster's larval experience is provided by M. F. K. Fisher: "It is to be hoped, sentimentally at least, that the spat—*our* spat—enjoys himself. Those two weeks are his one taste of vagabondage, of devil-may-care free roaming. And even they are not quite free, for during all his youth he is busy growing a strong foot and a large supply of sticky cementlike stuff" (Fisher [1941] 1988, 4).

The setting of oysters is a very critical stage in the transition from a mobile organism to one that is destined to be sessile, permanently fixed for the rest of its existence (unless it is disturbed by acts of nature such as severe storms or by human intervention). In northern waters, July is the peak month for oyster setting. Most larvae die at this stage because they don't find a proper attachment site. A drop in water temperature or an early summer storm can wipe out a whole season's production of oyster larvae.

The foot of the oyster larva shops around for a suitable home. As the larva assesses its substrate, it will make an important "decision." If it attaches to a substrate, it will be home forevermore. In contrast to clams and scallops, which produce a byssal thread that provides temporary attachment, the oyster will produce a bit of biological waterproof, cement-like adhesive that will set the oyster larva to its permanent home. At the time of setting, larval oysters are called "spat," because spawning was referred to "as spitting and they had been spat" (Kurlansky 2006, 119). Oyster spat are tiny specks barely visible to the human eye and it will take months before they noticeably resemble tiny oysters. The left valve, which will develop into the cupped or deeper shell, is the one that fixes to the site of attachment. After setting, the foot and eyespots are no longer functional; they disappear and the oyster can now feed and grow in place.

Biologists have speculated that oyster larvae will set where other larvae have set or that they set in clusters. This phenomenon has been tagged "gregariousness" and is hypothesized to be mediated by a pheromone released by recently attached spat (Hedeen 1986; Kennedy 1996). The idea of gregariousness is controversial; some scientists argue that it would be of questionable value to oysters if spat somehow tended to settle near other spat. This situation would lead to increased competition for food and space as the oysters grow. Another observation that might explain why oysters tend to be attracted to setting sites near other oysters is the finding that ammonia is an attractant for oyster larvae (Fitt and Coon 1992). With ammonia as the major excretory product of oysters and other bivalves, it follows that larval oysters might be drawn to areas where adult oysters are established successfully because these sites are likely to have plentiful food, optimal environmental conditions, and future possibilities for successful reproduction.

One problem that would confront oyster larvae if they are attracted to assemblages of adult oysters, such as in reefs, is the danger of being eaten by the adults. In laboratory experiments, it has been shown that 91.3 percent of oyster larvae do not survive a visit to the pallial cavity of adult oysters. However, the same set of laboratory experiments demonstrated that the incoming current produced by the cilia of feeding adults was not strong enough to capture a significant percentage of larvae (Tamburri, Zimmer, and Zimmer 2007), which would suggest that the slight chance of being cannibalized by adults is offset by the benefits of a communal living arrangement.

The setting of oysters is a key period during development and has been studied intensively. As opposed to clams, which burrow under the bottom, oysters do not have siphons and must stay on top of the sediment. Burial in sediment would be the death knell for an oyster, so setting allows it to stay on top of the sometimes mucky bottom. The setting of oyster upon oyster is the process that led to the formation of the expansive oyster reefs that once dotted the Atlantic shoreline. Charles Dickens apparently did not know about the biology of oysters and the fact that they live in large aggregations when he wrote *A Christmas Carol*. Dickens (1843) described Scrooge as "a squeezing, wrenching, grasping, scraping, clutching, covetous old sinner! Hard and sharp as a flint, from which no steel had ever struck out generous fire, secret, and self-contained, and solitary as an oyster."

By understanding the natural course of bivalve reproductive biology, scientists and culturists have made the leap from coastal bay, sound, pond, or estuary to the hatchery where the course of natural reproductive events can be manipulated or simulated under controlled laboratory conditions. The modern hatchery, can be anything from a waterfront shed using easily obtained supplies such as baking dishes and homemade equipment amid a myriad of pumps and PVC pipe to a high-tech laboratory. Whatever its condition, the hatchery is the backbone of the shellfish aquaculture industry— most shellfish culture operations are completely dependent on hatchery-produced seed.

Chapter 7

Emergence in the Hatchery

The replenishment of bivalve shellfish stocks in the wild by natural reproductive events is not a reliable process in the Northeast. In many areas, adult bivalves that serve as native broodstock have been overharvested or have become the victims of diseases that have decimated the population. In some areas where mature adults are present, larval production is not consistent from year to year due to poorly understood fluctuations in environmental parameters. In order to stay in the shellfish farming business, growers need a dependable supply of seed and thus they depend heavily on the efforts of shellfish hatcheries. A number of resources are available to guide individuals or companies that want to break into the seed bivalve–producing industry, ranging from "how-to" manuals on phytoplankton culture to a complete overview of the operation such as the guidelines produced by the United Nations Food and Agriculture Organization (FAO) (Baptist, Meritt, and Webster 1993; Helm and Bourne 2004).

During the depths of winter, when ambient seawater temperatures are plummeting, the water is warming up in the hatchery. A select number of individual bivalves are destined to produce the next generation of larvae to satisfy the seed requirements of shellfish farmers and to meet the needs of shellfish restoration programs. The parent bivalve cohort has been chosen carefully after consideration of specific traits such as rate of growth, survival, quality of meat, and ability to resist disease. Individuals from populations that exhibit these traits are further selected based on their overall health status and their size or age to ensure that they are mature. Some broodstock are selected on the basis of morphological traits, such as shell color or markings. For example, the shellfish genetics laboratory at the U.S. Marine Fisheries Service in Milford, Connecticut, has bred "skunk" bay scallops that have black and white stripes radiating from the hinge. These scallops will be useful as markers in restoration efforts because they will allow Long Island Sound harvesters to distinguish between natural populations and scallops raised in the laboratory.

The artificial breeding programs conducted by hatcheries inevitably result in loss of genetic variation because the broodstock only represents a small subset of the population. Continued breeding of the progeny of the broodstock will decrease the genetic variation even further. For aquaculture purposes, genetic variation may not be as important as selection for specific traits that will result in increased profit for the grower. However, if the bivalves are bred for restoration initiatives, hatcheries attempt to use as many individuals as broodstock as feasible, and they employ a number of strategies to breed or cross their strains. Despite these precautions, breeding programs may limit severely the "effective size" of the population, a term used by ecologists to describe how many individuals in a population actually are contributing to the gene pool. Even if a population is very large, inbreeding can occur if only a small percentage of the population provides the eggs and sperm that produce the next generation.

Notwithstanding the limitations of artificial breeding in terms of maintaining genetic diversity, hatchery programs have been very important in reliably supplying seed stock to shellfish farmers. Furthermore, these programs are key components of some important ecological restoration efforts in which no naturally occurring population is present in sufficient size or in suitable locations to repopulate an area that has been depleted of shellfish as a result of overharvest, disease, or other factors.

Phytoplankton Culture

In order to breed bivalves, a ready source of food must be available. In commercial hatcheries, about 40 percent of the cost of the operation may be used to produce enough food for the broodstock and for the larval forms of the molluscs. At Roger Williams University, in Bristol, Rhode Island, a steady cadre of students are employed to assist with hatchery operations. This university facility, the only shellfish hatchery in the state of Rhode Island, hums with bivalve-related activity during the winter under the direction of Assistant Professor Dale Leavitt. Troops of students work as caretakers for the hatchery and are involved in the myriad activities that are required to run a successful operation, from growing phytoplankton to producing and raising seed bivalves that are used for an oyster gardening program and restoration activities. Leavitt was formerly the Aquaculture Extension Agent in Massachusetts but was appointed to the same post in Rhode Island in 2003. He shares his expertise with shellfish farmers and teaches a semester-long course on "Practical Shellfish Farming." As a student in the course, I was able to witness hatchery operations first-hand and follow the winter/spring cycle of bivalve seed production. No matter how much science is involved in the process, quite a bit of artistry is essential in running a successful hatchery operation. Many years of experience have given Leavitt a feeling for the bivalves and the nuances in reproductive events and larval rearing.

Tubes, flasks, and jugs with contents in all shades of gold, green, and brown line the shelves of the laboratory section of the hatchery (plate 15). These flasks contain cultures of various species of single-celled algae, 2 to 10 μm (0.00008 to 0.0004 inches) in diameter, that will be mass-produced to feed broodstock as well as the resulting ravenous larvae. The algal species have been selected carefully for their cell size and nutrient content. At least one species is a diatom, beautiful creatures with delicate glassine shells made of silica. Over twenty types of unicellular algae have been selected to feed bivalves, but hatcheries routinely focus on five or six species that are easiest to grow and produce the best results for broodstock conditioning and larval growth. Pure cultures of the most desirable species can be purchased from commercial aquaculture suppliers or obtained from academic or government research laboratories. Algae are grown in clean, filtered or sterilized seawater to which nutrients such as nitrates, phosphates, silicon (for diatoms), trace metals, and minerals are added. Many formulas have been developed for nutrient media to feed algae and they are all similar in composition to synthetic plant fertilizers. Hatcheries once were required to prepare their own batches of algae food, but such mixtures, including the commonly used f/2 media and derivatives devised by Guillard, are now available from aquaculture suppliers (Guillard 1983).

The cultures are grown at the optimal temperature for the particular species, with an appropriate light source (fluorescent tube lights or grow-lights used in greenhouses) that allows photosynthesis, the mechanism that algae use to produce their own food. The smallest bivalve larvae will be able to filter only the smaller algal species but the size of their food particles can increase as the larvae grow. Among the popular algal species are naked flagellates, such as *Chroomonas salina* and *Isochrysis galbana*, nicknamed Tahitian Iso or T-iso, which are easy for small larvae to digest because they do not produce cell walls; green flagellates such as *Chlamydomonas coccoides* and *Nannochloris occulata;* greenish flagellates such as *Tetraselmis suecica* and *Pyramimonas virginica;* brown algae such as *Monochrisis (Pavlova) lutheri;* and diatoms such *as Skeletonema costratum* and *Chaetoceros calcitrans.* Growth of the phytoplankton is scaled up in a stepwise manner. Algal cells from the stock cultures are inoculated progressively into larger cultures. At Roger Williams University, the cultures go from stocks to 100 ml starter flasks to intermediate size cultures in 3.8-liter (1-gallon) containers (old wine jugs are employed after they have been scrupulously cleaned and sterilized), then to 19-liter (5-gallon) jugs. These, in turn, are used to start even larger batches of algae that will be produced in the greenhouse portion of the facility in large, clear-plastic, 200-liter (53-gallon) cylinders known as KalWall or Sunlite tubes. The cultures require one to two weeks of growth at each step of the process. The smaller cultures are static; that is, once the algae have grown, the entire contents will be poured into a larger-size culture vessel. The larger cultures are grown in a continuous system: When some of the phytoplankton are drawn off to feed the bivalves, more seawater is added and the remaining cells will reproduce further. Some commercial firms are now selling concentrated phytoplankton as microalgal pastes that can be stored frozen and diluted with seawater before use. These pastes provide a convenient supply of algae and are a good backup in the case of failed production in the hatchery.

Broodstock Conditioning

In nature, bivalves usually begin to release gametes in late spring, but in order to produce seed for spring planting, hatchery spawning must be induced during winter. Environmental conditions must be manipulated so that broodstock perceive appropriate cues for their gonads to ripen and for eggs and sperm to be released under controlled conditions. Leavitt uses stimuli that mimic natural spring conditions in the Northeast: fluctuating water temperatures, increasing concentrations of phytoplankton, and exposure (which would occur during tidal cycles). Once the hatchery has geared up the production of algae and has a good supply of phytoplankton on hand, conditioning can begin. The water temperature in broodstock tanks is increased gradually from ambient temperature of the water from which the broodstock have been obtained to temperatures that are required for the completion of gonadal development. The Roger Williams hatchery aims for an increase of 1 to 2°C (1.4 to 2.8°F) per day to achieve a maintenance temperature of about 20°C (68°F). At the same time that the water is warming, phytoplankton are introduced to the broodstock. The bivalves can be batch-fed periodically or fed smaller amounts continuously throughout the period. Quahog broodstock are feed 1 to 3 billion algal cells per animal per day! Water must be kept clean and waste must be removed. To

assess the success of conditioning and the reproductive status of the broodstock, it is often necessary to sacrifice a few animals for gross examination
of the gonads and for microscopic observation of gonadal tissues for developed eggs and sperm. The conditioning period and parameters vary somewhat with each species. For example, razor clams will condition best if they
are allowed to burrow into sediment. In general, it can take one to two
months of conditioning before broodstock are ready to spawn.

Spawning

When gonads are ripe and the hatchery has algal production sufficiently
geared up to feed larvae, spawning can be initiated. For most northeastern
bivalve species, the most common method to induce spawning is to spike
the temperature. The broodstock animals first are cleaned with a quick bath
in dilute bleach to rid them of bacteria that could contaminate future stages
of the process. The animals are kept in the cold, out of the water for 24
hours. In many hatcheries, the bivalves are placed in a spawning tray with
a dark background that allows a visual detection of the spawning process.
Some hatcheries will induce batch spawning, with all the animals initially
in the same tray; others prefer to keep the animals separated into their own
individual, clear-glass baking trays. A rapid rise in temperature serves as a
spawning trigger and once a single animal spawns, others follow suit. Temperatures are raised to 25 to 28°C (77 to 82.4°F) for 15 minutes; if spawning does not occur, the water is brought back to 20°C (68°F) and then
cycled back up to 25 to 28°C (77 to 82.4°F). If the rapid increase in temperature does not accomplish the job within a given time span, the temperature can be lowered and the temperature spike can be repeated. I once
watched an unsuccessful attempt to spawn conditioned bay scallops at the
Eastham Aquaculture Center on Cape Cod. The scallops would not coop-
erate but clearly were stimulated by the temperature spike. They treated
onlookers to a shellfish ballet as they clapped their valves, swirled in grace-
ful pirouettes, and fluttered about the tank. Temperature spiking works well
in northern species but is not effective in bivalves obtained from southern regions where water temperatures remain warm year-round.

If the broodstock is not cooperative, other stimuli might be employed.
For example, some hatcheries may add sperm, taken from a male in a process known as "stripping," or hormones such as serotonin. If all spawning
stimuli fail, gametes can be taken or stripped from males and females and
mixed together in a separate container of seawater.

In a batch-spawning tank, animals are removed immediately and placed in separate containers as soon as they begin to spawn and their gametes are identified as eggs or sperm. Sperm are emitted in wispy, milky streams, while the larger and grainier eggs are ejected in rhythmic waves, like puffs of smoke, and tend to sink to the bottom of the container. Separation of spawning animals prevents chance encounters between gametes so that fertilization can occur under controlled conditions. Eggs are pooled and the concentration of eggs is determined by microscopy using a special counting chamber; sperm is pooled and the concentration of sperm are similarly determined. It is important to know the concentration of eggs and sperm so that they can be mixed in optimal proportions for fertilization in order to prevent the fertilization of an egg by more than one sperm, a phenomenon known as polyspermy that can be responsible for abnormal larval development. The optimal ratio for fertilization in the laboratory is about 500 to 5,000 sperm per egg (Gosling 2003). Gametes collected in the laboratory or hatchery at high densities do not have a long shelf life. Once released, they must be mixed as soon as possible, preferably within the hour of their release. In hermaphrodites such as bay scallops, sperm usually are released first, so after sperm are shed, the bay scallop is transferred to a fresh container to collect its eggs and thus prevent self-fertilization.

Soon after fertilization, the eggs begin to divide. Microscopic examination of the samples from the fertilization mixtures is critical in order to determine the extent and success of the process. If a low percentage of eggs has been fertilized, more sperm can be added. Once a high percentage of eggs has been fertilized, as evidenced by ensuing cell division, they are washed of remaining sperm and placed in a tank with clean, filtered seawater. At this stage, the fertilized eggs become zygotes and begin the path of development into larvae.

Larvae

The most critical stage of seed production now begins, as the fertilized eggs continue to divide, develop into larvae, and undergo setting. The seawater-filled tanks are heated and aerated. During the first 24 hours after fertilization, the zygote develops into a trochophore larvae. During this time period, larvae do not feed; all their energy comes from maternal reserves in the original eggs. After 24 hours, the larvae have developed a mouth and gut and are ready and eager for food. Small algal cells must be supplied and can be delivered continuously or in several batches per day. With five to ten larvae per milliliter of culture, 50,000 algal cells per milliliter are required. The diet is doubled when the larvae are one week old and progressively increased as the larvae grow (Baptist, Meritt, and Webster 1993).

This is a critical period during hatchery operations. Tanks and water must be kept scrupulously clean to prevent contamination by bacteria such as *Vibrio* that have been known to destroy entire batches of larvae overnight. At Roger Williams, the larval tanks are drained and cleaned every other day. Because the larvae are planktonic at this stage (swimming and floating about), fine-mesh screens are used to collect them as the water is drained from their 250-gallon larval-rearing tanks. The small, transparent larvae are monitored daily for overall health, for signs of intact structures such as their velum, and for evidence of food in their gut. Larval caretakers also look for signs of metamorphosis that signal the competency of the larvae to set: the development of a foot, eyespots, and gill buds. Experienced hatchery operators note signs of "ropiness," evidence that the larvae are beginning to cling together into stringy masses, a macroscopic sign that they are ready for setting.

Larval Setting

The hatchery must be prepared when the larvae are about to set, one to two weeks after fertilization. Oysters, scallops, and quahogs set differently in the wild and these differences are reflected in the setting systems used in hatcheries. In nature, quahog larvae will explore the bottom with their foot until they find a suitable setting site. Without knowledge about the nature of the setting cues, it is not possible to mimic these conditions in the hatchery. Therefore, quahog larvae usually are collected on fine-mesh screens once they show signs of setting. Bay scallops and blue mussels will use their byssal threads to attach to substrates in the wild. Aquaculturists have developed synthetic substrates such as ropes or fine-mesh plastic to which larvae of these species will fix themselves. In the wild, oysters will use their byssal threads initially to attach to substrates, but the byssus will be lost as the oyster begins to produce the liquid cement that will fix it to its permanent home. In the hatchery, a hard substrate must be provided and substances such as whole shell, ground shell, or artificial cultch made from marble, plastic, or other materials can be employed. If the seed is being prepared for a shellfish farmer who plans on growing single oysters for the half-shell market, the best material is microcultch: ground oyster shells or even egg shells can be used. These particles have been ground to a very small size so they provide room for only a single oyster to set. If the oysters are being produced for distribution on the bottom, better survival is obtained if the larvae are set under conditions that mimic those found in the wild, so larger shells or pieces of shell are used. Due to the difficulty of securing sufficient oyster shell for cultching, shells of ocean quahogs and surf clams often are used as a substitute setting surface. Whole-shell cultch provides recesses for the larvae to hide, makes the larvae and young oysters more immune to predation, and prevents them from sinking into the sediments.

For some oyster farming operations and restoration projects in the Northeast, larvae are set in one location—sometimes a hatchery, sometimes in the wild if broodstock are available—and then moved to another location for growing. This practice is known as remote setting and the "set" must be in a form that can be transported. A common practice involves the use of mesh bags that are filled with whole shells or large pieces of shell that are put into setting tanks or placed in locations where spat might be collected. Once oyster larvae have set on this material, the bags can be moved to growing areas where the spat-on-shell can be dispersed.

Next Steps

Bivalves must be fed increasing amount of food during the setting period. Once a supply of young, larval bivalves has been produced, reach competency, and have "set" or been collected on mesh screens, the next step is the grow-out phase. Hatcheries do not find it cost-effective to produce enough algae to support the rapid growth capability of bivalves at this stage. The best food usually is found in the water where shellfish are found naturally. The most efficient and economical way to grow the hatchery-reared bivalves is to get the animals back into the water and allow them to sample the diversity of naturally available food. As water warms and phytoplankton levels increase, spring is the perfect time of year to let nature take over some of the work of bivalve rearing. Shellfish farmers are eager to collect the seed that they ordered from commercial hatcheries and plant them in their shellfish nurseries. As soon as environmental conditions are favorable, the outdoor work of the shellfish grower can begin. The next chapter describes this return to the great outdoors.

Chapter 8

Gearing Up and Re-Emersion

After the danger of ice has passed, shellfish farmers are able to return their pitted oysters to their "grants" where the animals will be in a more natural environment. Even though the water may be cold, the oysters will be happier than when they were sunken in the cellar or pit. By early spring, it is time for the growers to assess their winter losses of animals and equipment. Early spring also affords the perfect opportunity for shellfish growers to practice some public relations; in Wellfleet, this means the organization of a massive clean-up of beaches and marshland near shellfish growing areas. The Department of Public Works places a large dumpster in the parking lot at Mayo Beach on the date of the Annual Beach Clean-Up. Shellfish farmers and townspeople gather early in the morning, steaming cups of coffee in hand, to receive their assignments. With two large garbage bags, my husband and I headed to a cove where a large pile of debris accumulated over the winter, but soon realized that we would never be able to carry all the trash back to our car. Anticipating our dilemma, a shellfisherman soon arrived with his pick-up truck, which we quickly filled and he then relayed the load to Mayo Beach. By the end of the day, the dumpster was overflowing with all manner of human-generated debris. There were nets and bags used by shellfish farmers but also pieces of old boats and lots of plastic, rubber, styrofoam, and glass. Thanks to the shellfish growers, the coast always looks pristine after this annual event.

A bit later in the spring, when the water has warmed sufficiently, growers receive their seed bivalves, ordered months ahead of time from hatcheries. Small shellfish are vulnerable to predators and must be grown to a larger size before they can be placed into typical grow-out containers or placed on the bottom. Tiny bivalves, resembling grains of sand, are ushered through a nursery stage to give them a head start in growth. All manner of techniques and equipment are used for shellfish nurseries, but the simplest method is to create a nursery area within the shellfish farm itself. I accompanied my neighbor Russ to his farm one day to tend his oyster nursery.

Two weeks prior to my visit, Russ had received his seed oyster from two different commercial hatcheries. When it comes to oyster seed, farmers have learned not to put all their eggs in one basket; in any given year, seed from one hatchery may outperform seed from another hatchery, so it's better to diversity when purchasing seed. When we waded to the "grant," Russ remarked that his seed, contained in fine mesh bags within larger plastic mesh bags, and the size of fine gravel, already had doubled in size. The oysters needed to be spread out and given more room to grow. They were large enough to be retained on the typical window screening that Russ used in the construction of his nursery-tray system. My job was to transfer the seed oysters from the bags to the nursery trays, being careful not to lose a single specimen. It was obvious that Russ's nursery was in an ideal location, providing enough food and suitable environmental conditions to support the rapid growth of his seed oysters. Russ would check the nursery trays periodically, and within a few weeks, the oysters were destined to be separated again to prevent crowding and competition for food.

Nursery Systems

Not all shellfish farms have the same conditions as those found on Russ's "grant." Furthermore, if an aquaculture operation includes seed scallops and quahogs, it requires nursery conditions that are different from those used for oysters. The type of nursery chosen by a shellfish grower depends on many factors, including the type of bivalve being cultured; whether the site is intertidal, subtidal, on a dock, on a barge, or on land; as well as availability of space in a suitable geographic location, water quality, environmental factors, and cost. Each grower has a preferred system that sometimes is chosen after trial and error or observing what works for neighboring culturists. Nursery sytems can be as diverse as the growers themselves; often, the successful nursery is the product of ingenuity, intuition, and the grower's experience.

Field Nurseries

In areas where natural populations of adult oysters give rise to planktonic larvae in the water column, shellfish farmers sometimes opt to catch a natural set, that is, to collect and nurture their own seed, rather than purchase it from hatcheries. My neighbor, Paul, uses devices known as Chinese hats to entice the settlement of larval oysters. Chinese hats are shallow plastic cones that can be stacked atop one another. Paul mixes up a batch of a special coating containing cement, sand, lime, and water, and dips the stack of Chinese hats into the thin, creamy slurry. Shellfish farmers often have their own secret recipes for the exact proportions of the components of the cement. The approximately 1-meter (3 to 4-feet) tall stacks of Chinese hats are set out in areas that are known to collect a good set of oysters and placed into the water column just before the reproductive season, (plate 16). In an interesting interpretation of this technology, the use of Chinese hats is considered to be fishing, rather than shellfish aquaculture, so the hats do not have to meet specific height restrictions that are imposed on other shellfish culture technology. (For example, the U.S. Army Corps of Engineers mandates that most aquaculture gear extend no more than 0.46 meters (18 inch) from the bottom so that it does not present a navigation hazard.) If the farmer is lucky and obtains a good set on the hats, he will have his own supply of oyster seed and will not need to purchase seed from a commercial hatchery. Once the seed oysters have grown to a suitable size, they can be removed by flexing each disk to flake them off as single specimens. Paul prefers catching wild spat because he feels it represents the progeny of oysters that have successfully grown and reproduced in the local waters. Other farmers start from scratch and catch wild spat because it's free for the collecting. Even though preparing, carrying, and placing Chinese hats is labor-intensive, this method will increase a farmer's profit margin because he or she doesn't have to invest in seed.

Field nurseries employing small mesh bags and racks such as those used by my neighbors Russ and Paul are the least-expensive systems for oysters. However, these intertidal systems are not appropriate for bay scallops, which do not tolerate dry conditions, nor quahogs, which prefer to bury in the sand. Quahog nurseries may consist of bottomlands in the intertidal that have been specially prepared and raked to remove predators, seeded with small quahogs at appropriate densities to prevent overcrowding, and covered with protective netting. These nurseries usually are set up as long rectangular plots, sometimes called "raceways (fig. 8.1). Modified field nurseries for quahogs are constructed from boxes or pens that are placed on the bottom and filled with sand. Some variations may have wire mesh on the bottom while plastic netting is used to cover the top. A variation of the quahog nursery design uses partially submerged boxes filled with sand, thus keeping the tiny bivalves off the bottom where crabs would feast on them. If the nursery trays are on the bottom in a subtidal area, the shellfish grower must have a system to lift the trays so they can be cleaned and inspected periodically.

FIG. 8.1 Field-planted clam raceways in Welllfeet. *Photo: Barbara Brennessel*

A variety of designs for floating field nurseries can be used in locations where there is plentiful food in the water and where local regulations allow the placement of floating, raft-like structures that might be tied to docks. One version, the Taylor Float (named after Jake Taylor, who designed the prototype), can be used both as nursery and as a grow-out system for several types of bivalves. This off-bottom system is a low-cost nursery alternative that can be home-made from readily available materials and provides protection from shellfish predators. The float measures about 0.6 meters (2 feet) wide, 2.4 meters (8 feet) long, and 0.3 meters (1 foot) deep (fig 8.2; Luckenbach and Taylor 1999). A 1-inch mesh covers the bottom but the float can be used for smaller oysters if they are placed in smaller mesh bags within the containment of the floating nursery. Because of possible navigational hazards and local regulations, Taylor Floats are not an option for growers in all areas.

Blue mussel nurseries usually are located in the field because it is the practice of mussel farmers to catch wild seed on artificial substrates or to transplant wild seed from locations that differ from the growing area. Larval blue mussels will attach to almost anything; oyster growers consider them a nuisance because they will attach to oyster-growing gear and compete with oysters for food. However, the ability of larval blue mussels to adhere to all manner of material is a boon for the mussel farmer who sets out ropes, plastic mesh tubing known as mussel socks, and other materials into the water column to attract mussel larvae. The mussel catchers usually are suspended from rafts or a system of ropes that hold the gear so it can float at the appropriate depth. Once the mussels have grown to a certain size,

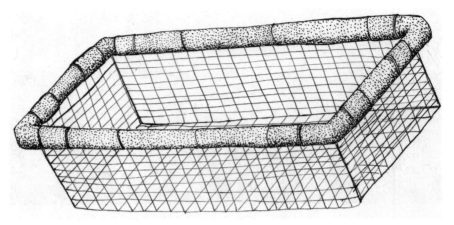

FIG. 8.2 Taylor float. *Drawing: Marisa Picariello*

they often are transferred to mussel socks with larger mesh and suspended on rafts, poles, or floating structures known as long-lines, for their final grow-out. Compared to bottom culture, mussels grown in suspended systems don't pick up grit and are less likely to form undesirable, misshapen pearls.

Softshell clams are more difficult to culture than other bivalves. Nurseries may consist simply of a clam tent that is placed in a sandy area in the intertidal (fig. 8.3). Seed clams purchased from a nursery can be buried and netted or the grower can rely on Mother Nature. Clam tents recruit and protect a natural set of softshell clams by attenuating the current around the net and proving a refuge from predators. Softshell clam larvae attach to

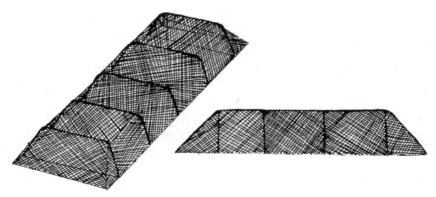

FIG. 8.3 Softshell clam tent. *Drawing by Peter Dion; courtesy of Coastal Aquaculture Supply*

sand grains in the sediment where they will burrow and grow. Shifting bottoms and the presence of predators can quickly eliminate the tiny clams, which may be less than 0.25 mm (0.01 inch) long.

Alternative Nursery Systems

Rather than relying on the tides or currents to provide food, some growers are more proactive and use systems to bring the food to the tiny bivalves. The water-based raceway described above is located on a grower's leased area, usually in the intertidal zone. A different type of raceway used as a nursery to grow shellfish seed is located on terra firma. Land-based raceways are onshore nurseries, commonly used by hatcheries and research labs, such as the U.S. Department of Marine Fisheries facility in Milford, Connecticut. The raceways consist of long, shallow trays made from concrete, wood, or fiberglass; all but the cement versions can be stacked on top of one another to maximize the use of space. Although they may be constructed using different materials and configuarations, they all use massive pumping systems to deliver large volumes of water to the seed. The seed is spread in a thin layer along the bottom of the long tray and then seawater is pumped into the system from a nearby source. The water enters either from one end via multiple ports, or is sprayed into the tray with a manifold, allowing more oxygenation. The water flows horizontally through the raceway, providing oxygen and nutrients to the bivalves, and then drains at one end and returns to the source. For clam culture, the floor of the raceway sometimes is covered with a layer of sand (Castagna 2001).

Another approach to nursery culture utilizes a device known as an upweller. Upwellers can be land-based or floating. Land-based systems usually are placed near the water so that they can draw water from a nearby source and return the water to the same source after it has nourished the bivalves. Some land-based systems are enclosed in buildings but others are open systems with removable coverings. For the land-based systems to work, water must be lifted from the sources by powerful pumps and therefore they are expensive to operate.

In an upweller system, the structures containing the bivalve seed, known as silos, are placed in a large trough while the water is moved by a pump from below the screened silos, upward through the bed of seed organisms, and exits from a pipe placed near the top. The pump is working constantly to empty the pipe that is pulling water from each silo containing bivalve mollusc seed. The water supplies food and oxygen and removes waste prod-

FIG 8.4 FLUPSY at Wellfleet town pier. *Photo: Barbara Brennessel*

ucts, thus allowing very dense concentrations of bivalves to be grown in a small, protected area that can be monitored easily. This type of nursery set up maximizes growth and minimizes mortality.

A practical configuration for an upweller that has become increasingly popular is a floating system. This system can combine the benefit of an upweller with a recreational or commercial dock. The upweller may be incorporated into the dock but is below the planking and not visible to marina visitors or patrons (fig. 8.4). The silos containing the seed bivalves are accessible through hatch doors on the deck. The nickname for this type of nursery system, which can be used for oyster, scallop, and clam seed, is "FLUPSY," for FLoating UPweller SYstem. Water is moved through the system with an airlift, similar to those used in aquaria, or a small marine pump, generally ½–¾ horsepower, which can force gallons of water through the silos each minute. The pumps are generally the same type as "Ice Eaters," submersible motors with small propellers that were developed as de-icers to prevent ice from forming around boats and piers and within marinas during the winter. Even though electricity is required, FLUPSYs are less expensive to operate than land-based upwellers. Furthermore, waterfront property is not required. Aquaculture specialists at Mook Sea Farm in Damariscotta, Maine, have experimented successfully with upwellers that are powered by tidal flow, making them even less costly to operate. Fully assembled upwellers can be purchased from aquaculture suppliers, but more often than not, shellfish growers will construct their own devices from materials available at building supply stores, as was done in Southold by the Cornell Cooperative Extension for their oyster propagation project (Rivara, Tetrault, and Patricio 2002).

Nursery Maintenance

No matter what type of nursery is utilized, bivalves must be checked periodically and the system must be maintained. A common and frequent problem is biofouling, the growth of unwanted marine organisms, which can clog up small-mesh screens and bags as well as pipes and prevent suitable water flow (described in chapter 14). During this period of rapid growth, some bivalves will outpace others, so that growers must continually sort the nursery stock by size, usually by sieving through screens with specific mesh sizes, and place similar-sized animals together. This prevents larger ones from consuming all the food and making it difficult for the smaller ones to grow.

Aside from providing conditions to support rapid growth, an additional advantage of shellfish nurseries is that they can be located in sites where shellfish harvest is not permitted because of water-quality issues. If there is enough food in the water, the tiny bivalves will grow rapidly and can then be moved to approved sites for final grow-out.

Chapter 9

Location, Location, Location
The Siting of a Shellfish Farm

With continued effort by the shellfish aquaculturist, bivalves (predominantly oysters, quahogs and scallops) that graduate from the shellfish nursery can be grown to sizes that permit harvest and sale. In some cases, the grow-out area is the same as the nursery that was used to nurture the seed from its collection as spat. In other instances, the seed is moved to a shellfish farm, which is an area of subtidal or intertidal bottomlands, licensed for aquaculture and commonly called a "grant." The procedures for granting of licensed shellfishing areas differ from state to state. Some states in the Northeast hold intertidal and subtidal areas in trust for state residents and the location of shellfish farms is overseen at the state level. In contrast, Massachusetts and Maine continue to abide by colonial ordinances dating back from 1641 to 1647 and honor individual deeds to tidal lands while designating jurisdiction and oversight of shellfish aquaculture areas to individual towns.

Before an individual or company initiates the permitting process for a shellfish aquaculture location, they have to do their homework and consider many variables in selecting a site for growing shellfish. One of the most important parameters is to avoid conflicts with shellfishermen who pick, rake, or dredge natural stocks of shellfish. Site selection necessitates the identification of locations without naturally occurring shellfish, that is, barren or unproductive areas. The location also should be one in which the farming operation will not have a negative impact on a threatened or endangered species. For example, eelgrass (*Zostera marina*) is protected in most cases and shellfish farms cannot be sited on beds of eelgrass. The potential aquaculturist also must do some research on the species of bivalve that will be grown. Scallops cannot tolerate being out of water for long periods, so an intertidal location would not be a good choice for a scallop farm. Cold-water bivalves, such as surf clams, would not survive on warm intertidal flats. The potential site should be assessed for expected winter conditions and the degree of icing. If the grower plans to use bottom-culture techniques, the nature of the substrate should be evaluated. For example, it would not be very wise to use bottom culture for an infaunal species, such as clams, on very rocky substrate, nor is it wise to culture oysters on a bottom location that is subject to frequent siltation, which can result in smothering of the bivalves.

When the overarching concerns about site selection have been addressed, other parameters factor into the siting equation: good water quality, appropriate environmental conditions such as water temperature and salinity, and plenty of phytoplankton that the bivalves can use as food. Ideally, the site would not be near areas of agricultural runoff or receive large influxes of freshwater that can suddenly or dramatically decrease salinity or increase pollution from land-based sources. The shellfish culturist's homework should include the study of potential predators and pests, so that there will not be major surprises once an investment in the farm has been made. Another consideration for growers is access to the site as well as to their gear and equipment, because it is critical that the shellfish farm be tended regularly. Different modes of access are necessary when the grant is on the intertidal as opposed to subtidal, so the farmer must know how he or she can reach the grant by foot, truck, or boat.

Once a possible site is selected, the permitting process can be initiated. Each of the Northeast states has its own procedural guidelines for applying for a shellfish aquaculture "grant." The grower essentially leases the bottom or is granted permission to use the area from the state or municipality (hence the term "grant"). Aquaculturists obtain a license for exclusive use of the bottom for the purpose of cultivating shellfish. They own their gear and equipment as well as all shellfish grown on the "grant," but they usually don't own the bottom unless they have title to the property. Each northeast state has developed specific goals and approaches to guide the development and implementation of bivalve shellfish aquaculture and enforces its unique set of rules and regulations. The following state-by-state survey gives an overview of the similarities and differences among the states.

Maine

In Maine, shorefront property owners have title to the bottomland down to the mean low-tide line, but not farther seaward of the mean high-water mark than 100 rods or 1650 feet (503 meters). This is a holdover from the the colonial era, when Maine was part of Massachusetts Bay Colony, which enacted the Colonial Ordinance of 1641–1647 to encourage wharf building and commerce. This colonial ordinance is still binding because Maine retained Massachusetts Common Law after it earned statehood. Despite somewhat restrictive ordinances, a leasing program for shellfish aquaculture has been in place in Maine since 1971. Main is home to about 70 shellfish farms, growing mostly Eastern oysters, European flat oysters, quahogs, blue mussels, and softshell clams. Some farmers are experimenting with surf clams, razor clams, sea scallops, and bay scallops. The farms are scattered throughout the state, with concentrations in the Damariscotta River (a historically rich shellfish area), Casco Bay, the St. George River, Sheepscot River, and Penobscott Bay.

In Maine, a propective shellfish farmer can take two different courses to obtain a "grant," depending on the proposed location. One leasing program for cultivation of marine resources in intertidal and subtidal locations is overseen by the Maine Department of Natural Resources (MDNR). Leases are valid for ten years with annual rental fees. Each application is limited to 5 acres of bottomland and each grower is limited to a total of 100 acres. The leases cannot be sold or transferred to other individuals. Applicants must demonstrate that their shellfish farm or its operation is not within 1,000 feet of a government-owned beach, park, or docking facility, does not interfere with other coastal uses such as navigation and fishing, and does not prevent access to the water by upland property owners. If access to the site is via private property, the property owner's permission must be secured. The initial application to MDNR is followed by a site visit by a biologist, input from the Coast Guard, Army Corps of Engineers, Maine Department of Environmental Protection, the State Bureau of Public Lands, and the State Planning Office. After passing these hurdles, the application must undergo a public notification process and subsequent public hearing. When all stakeholders are on board with the project, a final review is made by the Marine Advisory Council, which then makes a recommendation to the MDNR Commissioner, who is responsible for the final approval. These leases are most easily obtained by shorefront property owners, as well as other fishermen who traditionally have worked in the area.

Towns in Maine also have jurisdiction over intertidal shellfish leasing programs, provided that the municipality has filed an approved shellfish conservation program. After an application is made for a shellfish farm, an assessment is made of its impact on the town's shellfish conservation program. There is a public notification process, a comment period, and sometimes a public hearing. Size restrictions, as well as compliance with state regulations, also apply to municipal permits but municipalities can impose additional restrictions. One might wonder why permitting a shellfish farm can be part of a town's shellfish conservation program, but some potential benefits may be gained by this type of arrangement. When cultured bivalves spawn, gametes are broadcast into the local water body and this reproductive gift can augment natural populations of bivalves. Furthermore, if a shellfisherman is culturing bivalves, he or she is less likely to be depleting the town's natural stocks.

No matter which program provided the grant, the shellfish culturist must obtain additional permits to import shellfish, to sell and/or process shellfish, and in some cases, must have an Interstate Transportation License if the product is destined to leave Maine.

New Hampshire

Shellfish aquaculture in New Hampshire is in the experimental phase and not yet a major industry. Some oyster culture is being conducted for restoration purposes and limited recreational, noncommercial harvest of oysters is done by tonging. In addition, the University of New Hampshire has initiated an Open Ocean Aquaculture Demonstration Project, with blue mussels as the test species, to assess the technical and economic feasibility of open-ocean shellfish culture. Using techniques that have been employed in New Zealand, Canada, and Europe, the potential success of this project may provide a direction for the future of shellfish aquaculture throughout the region.

Massachusetts

In Massachusetts, "Home Rule" rules. Each town has oversight of shellfish aquaculture operations and can issue shellfish aquaculture licences after public notice is filed, a hearing is held, and certification is obtained from the Massachusetts Department of Marine Fisheries under Massachusetts General Laws, Chapter 130, section 57. The exception occurs when there is a risk to human health, which allows the State to shut down certain areas for shellfishing and shellfish farming. The Army Corps of Engineers, local conservation commissions, and other agencies are always involved, but in most cases, this is a town-by-town process. In general, those applying for leases must be residents of the town where the farm will be located and leases cannot be sold or transferred. Furthermore, towns cannot issue leases on intertidal lands that belong to upland property owners. Even though the colonial ordinances grant the public the right to use private intertidal land for "fishing, fowling and navigation," with shellfishing included under the umbrella of fishing, towns cannot grant permission to cultivate shellfish on privately held property. This was made clear by the Supreme Judicial Court of Massachusetts in the now-famous "Pazolt decision," rendered when a lawsuit was filed after the town of Truro granted a shellfish license on flats to which a motel owner held title. In essence, the ruling stated that shellfish aquaculture is not a natural derivative of the public's right to fish primarily because it involves placing structures onto the bottom rather than just digging or collecting. In other words, instead of fishing, shellfish aquaculture is more akin to agriculture and a town cannot give a person a lease to farm on someone else's private property.

Some Massachusetts towns have performance requirements that must be met in order for a grower to keep his or her shellfish aquaculture license. For example, Mashpee requires that a shellfish farmer plant a minimum of 110,000 shellfish per year. The major growing areas are in the Cape Cod towns of Wellfleet, Eastham, Dennis, Brewster, and Barnstable. In addition, shellfish culture occurs in areas within Pleasant Bay, such as Orleans and Chatham, towns along Nantucket Sound such as Cotuit and Mashpee, the islands of Martha's Vineyard and Nantucket, across Cape Cod Bay in Duxbury, and locations along the North Shore (north of Boston), some of which are famous for softshell clams.

Rhode Island

In Rhode Island, one is never far from the water. Providence was once the center of the state's oyster industry, with the main beds located in Narragansett Bay. Overharvesting, pollution, and the Hurricane of 1938 took their toll on the natural oyster beds. In the 1940s, after the decline of the oyster industry, quahogs became the major focus of the shellfishery. Quahogs remain closely associated with Rhode Island's identity. A popular animated television show, *The Family Guy,* is set in the fictional town of Quahog, while the main adult characters take their libations at an establishment called The Drunken Clam.

For many years, commercial fisherman of wild shellfish were opposed to shellfish aquaculture because they felt that shellfish aquaculture privatizes bottomlands that are considered free and common to all. However, with declining wild stocks, it's no surprise that the shellfish aquaculture industry presently is considered as an avenue for economic development. With support from the legislature and almost 30 farms in operation, the industry is growing slowly. Shellfish, consisting mostly of oysters, are grown in various parts of Narragansett Bay and in the salt ponds that fringe the southern coast. In Rhode Island, the bottom, to the mean high-tide line, is owned by the state. The Coastal Resources Management Council (CRMC) provides oversight for state residents to obtain a shellfish aquaculture permit. When a grower has identified a potential shellfish farming site, the initial permit application is reviewed by the CRMC. The review involves an inspection of the site, public notification, public hearing, and evaluation by the governing board of the town where the site is located. Potential user conflicts must be identified and considered. This part of the process can be facilitated by consulting a series of charts published by the CRMC Working Group on Fisheries and Aquaculture and posted at http://www.narrbay .org, which map out the use of state waters by various interest groups including fishermen and boaters, as well as those areas under conservation protection. Once CRMC gives preliminary approval for the site license, the U.S. Army Corps of Engineers (ACOE) makes an assessment of the site. The ACOE always is involved if any structures are to be placed in the water, including aquaculture gear such as trays, boxes, cages, and even nets. If the ACOE gives their approval, the application goes back to the CRMC for final approval. The permission to lease is valid for ten years, with possibility of renewal, but the lease itself must be renewed on an annual basis. A farmer who receives a license must then apply for a permit for aquaculture from the Rhode Island Department of Environmental Management that will allow the "grant" holder to keep shellfish on the property.

The CRMC also has an abbreviated permitting process for individuals who want to test the feasibility of a certain area for shellfish cultivation. The applicant applies for a Commercial Viability Aquaculture Permit limited to 1,000 square feet. The grower has a three-year period to find out if the site is workable and whether shellfish will be able to grow.

Connecticut

In Connecticut, about 43 large-scale shellfish companies focus mostly on oysters and clams, with some blue mussel and razor clam experimentation (Getchis 2005). This activity is centered on about 70,000 acres in the Connecticut waters of Long Island Sound, where 674 franchise oyster grounds (more than 100 years old) utilize 22,422 acres, and 336 leased areas encompass 35,686 acres. In addition, municipalities lease about 12,000 acres. Currently, private oyster companies may move their animals several times, from areas where they set, to growing grounds that supply more food, and finally to clean, certified waters that meet national sanitation standards from which shellfish can be harvested.

Cultivated quahogs are collected from leased or franchised bottomlands with hydraulic dredges that soften the sediment and loosen the clams for easy harvest (Getchis 2006). Bay scallops are raised for restoration purposes. As in Rhode Island and Massachusetts, the shellfish industry is considered an important economic benefit to the state.

Similar to Maine, both a state and a municipal avenue are available for obtaining shellfish grants. An application process for leasing shellfish growing areas requires oversight by the ACOE, Connecticut Department of Environmental Protection, and municipal shellfish agencies. Harbor managers and conservation commissions also are involved. Prescribed procedures for application to lease state shellfish beds ultimately must be approved by the Connecticut Commissioner of Agriculture. The process for obtaining a grant from the state is a bit unusual because it involves a bidding process for a three-year grant at a minimum bid of $4 per acre. Unfortunately for those who want to break into the industry, no state shellfish bed leasing opportunities currently are available in Connecticut. However, small grants for individuals still may be obtained from certain coastal municipalities.

New York

In the 1970s, we harvested quahogs by walking out, waist deep, into Peconic Bay from the home of a friend, using our toes as clam rakes. The bottom was sandy and clean, devoid of sharp shells or rocks that might cut our feet. A half hour of toe-digging would produce enough large quahogs to make a white clam sauce to pour over linguini to feed family and friends. My husband makes the same sauce to this day. The key to the flavor is the freshness of the clams and all their juices, garlic, fresh parsley, and red pepper flakes. It's a good way to use chowder clams if you don't want to make clam chowder (see the recipes in the appendix).

Bay scallops were also abundant. If we were charged with procuring the evening meal, we might bring our masks and snorkels to the beach on Peconic Bay and dive for dinner. The bay scallops we collected were so sweet that any type of preparation would be devoured: No leftovers for this meal! One of our favorite ways to prepare them was to coat the scallop muscle with a bread-crumb or corn-meal mixture, liberally sprinkled with red pepper flakes, and then dip them into hot oil for a quick fry. Served with lemon wedges, the taste was "to-die-for."

Although there was a time when New York waters were teeming with shellfish and they were harvested in great quantities from the wild, this is not the case today. Shellfish farming is a relatively recent development that is being viewed as a technology to restore the fishery to its glory days.

The New York City area itself was so rich with oysters that the history of the city was written around this delicious mollusc (Kurlansky 2006). The long-standing shellfish industry in New York always has focused on the southern portion of the state, in areas around the Big Apple itself, such as in Raritan Bay, near Staten Island, in Jamaica Bay, Queens, and on the outstretched finger of Long Island, encompassing Nassau and Suffolk counties. Today, Raritan Bay shellfish are "polluted" and must be transplanted to cleaner waters, often in Suffolk County, before they can be sold.

Some of the shellfish growing areas in the state, such as the bottomlands in Great South Bay, between Fire Island and the mainland, were under franchise to large oyster companies since colonial times. By 1967, 50,000 acres of bottomland owned by New York State was leased or franchised for shellfish cultivation. Due to a number of factors, including lack of seed oysters, pollution resulting from industry and urban development, over-harvest, and disease, much of the leased oyster grounds became unproductive and were surrendered back to the state (Timmons et al. 2004). In 1983, laws were enacted that repealed the perpetual franchise system such that all new oyster growers were expected to apply for fixed-term, renewable leases. However, the industry remained at a standstill for a variety of reasons. For example, a long-lasting brown tide event contributed to the almost total disappearance of shellfish from Peconic and Gardiner's bays in the 1980s. In 1990, 80 percent of remaining Long Island Sound oysters were devastated by Dermo disease (see chapter 13), and the population has not yet recovered. Today, 1,694 acres of bottomland are still franchised within the New York state borders of Long Island Sound, and of approximately 45,000 acres that were franchised in Peconic Bay and Gardiner's Bay, all but 3,400 acres have reverted to the state (Timmons et al. 2004). A few of the major shellfish companies are still in operation, such as Frank M. Flower and Sons, which maintains a lease in Oyster Bay. The Bluepoints Company, which held title to over 13,000 acres in Great South Bay, recently went out of business and its bottomlands were acquired by the Nature Conservancy.

In 2004, Environmental Conservation Law 13-0302 was passed in New

York State in an effort to revive the once-thriving shellfish industry. This legislation, which has become known as the "2004 Leasing Law," granted approximately 100,000 acres of state-owned bottomland in Peconic and Gardiner's bays to Suffolk County for developing a shellfish leasing program. The county has until the end of 2010 to conduct surveys to identify appropriate shellfish cultivation zones, develop a leasing plan, and issue leases, or else the bottomlands will revert back to the state. The Suffolk County Aquaculture Lease Program Advisory Committee (ALPAC) has begun to prepare the necessary documents, such as a Draft Generic Environmental Impact Statement. The net result will be guidelines for streamlining the lease process to reflect current practices and to provide for managed growth of the industry.

Playing by the Rules

Although regulations and methodologies are different from state to state, some local oversight and enforcement of the rules must exist in order to protect the growers, other fishermen, citizens in general, and the environment in which the industry operates. I got a first-hand look at how a municipality oversees day-to-day shellfish aquaculture when I made morning rounds at low tide with Wellfleet Assistant Shellfish Warden John Mankevetch on a picture-perfect September day. John and other local shellfish wardens have to draw a fine line between helping out the shellfish farmers and wild pickers and enforcing local regulations. John's first checkpoint was at the rain gauge attached to the Shellfish Department building. Did I say building? It's more like a shellfish shack, situated at Mayo Beach, on the water's edge near the town pier. The rain gauge is checked because a big rain event may result in closure of certain shellfish areas. Large rainfalls are responsible for increasing the coliform bacteria levels in specific shellfishing areas in the Herring River. Without performing bacterial counts, shellfish areas are closed automatically in certain areas after heavy rains. After John checked the gauge and recorded the amount of rain (very little), we went to the next checkpoint: a video camera set up to observe poaching activity on a grower's "grant." The Wellfleet Shellfish Department installed the camera in an attempt to catch the culprit who was stealing a farmer's oysters; the camera was still operating but the tape showed no illegal activity. Next, we boarded the distinctive, forest-green, shellfish truck and headed to various landings to check on the folks who were "working the tide." When John had to drive across salty tidal flats to monitor all the low-tide activity, it became easy to understand why the Wellfleet Shellfish Department requests new trucks so frequently. The vehicle was showered in the saltwater that splashed from beneath the wheels. We started out in north Wellfleet and worked our way south, looping back to Route 6 and then heading west to get to the water multiple times. We stopped to check out the town beds, where quahogs are raised for the noncommercial harvesters, the recreational clammers. An important part of John's job is to grow 1-inch-wide, legal-sized quahogs for the "put and take" activity of recreational permit holders, necessary because there are no longer many littleneck clams to be found in Wellfeet outside of shellfish farms. John's efforts are appreciated by noncommercial shellfish permit holders, especially vacationers, who look forward to raking up dinner as an annual family activity. Their recreational permit fees subsidize the Shellfish Department budget and their interest in shellfishing in Wellfleet is thought to contribute to the local tourist economy.

One of John's duties is to listen to the "word on the tide" as he talks to the shellfish farmers when they make their way out to their "grants" or as they return with their trucks loaded with quahogs or oysters. On this particular day, John was on the lookout for shellfish aquaculture violators. For example, he had heard rumors that a farmer had moved his "grant" markers to take advantage of extra space on his neighbor's licensed site. When a shellfish aquaculturist is awarded a shellfish "grant," many towns mandate that the area be properly surveyed and marked off with buoys that delineate the perimeter of the shellfish farm and also warn boaters, windsurfers, jet skiers, and others that there is gear underwater. Wellfleet shellfish regulations dictate that all gear must be a maximum of 18 inches above the bottom. Most gear is exposed completely at low tide and partially exposed at intermediate tides, which creates a potential hazard for other users of the area. Thus, it is important that the shellfish farms be marked off and identified. If buoys are moved intentionally, the farmer may be claiming more farming area than originally granted. These types of boundary disputes must be mediated by the Shellfish Department and markers must be placed in their proper locations, determined by licensed surveyors and checked with GPS (Global Positioning System) coordinates. It's helpful to have diplomatic experience to solve these disputes, which sometimes are caused intentionally but more often occur accidentally. John also was checking out another shellfish farmer who was observed harvesting "shorts" and moving them onto his "grant." Shorts are oysters that are smaller than legal size. In Wellfleet, it is illegal to harvest wild oysters under 3 inches in height. If a farmer obtains "shorts" by dredging or picking wild oysters and moves them to his or her farm, other fisherman will not have the opportunity to harvest them in an area where they can continue to grow to legal size. Most Wellfleet shellfish farmers are honest, hard-working, and dedicated to their profession, but there are always a few who take advantage of a situation and keep the Shellfish Constables on their toes. Violators of shellfish harvest and farming regulations are subject to fines and license revocation. On this particular day, we discovered no signs of illegal activity.

We then moved on to check the productivity of a certain area of Wellfleet Harbor that had been devoid of oysters in recent years. Cultching operations by the Shellfish Department apparently were successful, because we observed oysters of all sizes in the area and a few shellfishermen were filling their baskets with legal, 3-inch oysters.

All towns that permit shellfish aquaculture on public lands have a system of oversight similar to the one that I observed while making rounds with John. The right to raise bivalves on state and town land is a privilege granted to the aquaculturist, and activities conducted at the farms must be lawful and enjoyed with certain important restrictions to protect wild fishermen, other shellfish culturists, and the public at large.

Chapter 10

Working the Tide

Grow-Out

All sorts of equipment and gear can be put to use in shellfish farming and most of it can be purchased from aquaculture supply companies. To see some of the gear in pristine condition, before it was subjected to saltwater and covered by masses of algae and other fouling organisms, I visited Coastal Aquacultural Supply in Cranston, Rhode Island, an outgrowth of a commercial pipe supply company. Owner Brian Bowes encourages site visits from marine educators and their students as a way to provide information about the potential economic benefits of shellfish cultivation projects and also to give students an idea about the future of shellfish farming. Bowes gave me the grand tour of the facility, which sells everything from clam rakes to FLUPSYs, with many items imported from around the globe. Bowes showed me "R9" polypropylene netting from Italy, Chinese hats from France, bags from Venezuela, France, and Spain, rope from Portugal, lantern nets from China, and trays from Australia and Canada. The type of gear purchased by each grower tends to be very individualized, with most growers trying a few different types of equipment to see what works best under conditions at their shellfish farms. Most states require that gear be labeled with the grower's name and contact information so the owner can be traced in case the equipment gets torn away by storms or is abandoned.

In addition to the type of equipment employed, the nature and techniques of cultivation depend on the species being cultivated, the size of the seed, and the parameters of the site. It is a well-known tenet of shellfish aquaculture that the larger the seed, the greater its probability of survival, but the greater the initial cost for the grower. Temperature requirements for cold-water species such as surf clams and sediment requirements for infaunal species such as quahogs and softshell clams will necessitate different growing conditions than those used for oysters.

FIG. 10.1 Mesh bags on oyster farm. *Photo: Captain Andrew Cummings*

Oyster Equipment

Oyster culturists use a variety of methods for growing their small oysters to market size. In deep-water sites, oysters that have graduated from the nursery can be cultivated directly on the bottom, where they can be harvested later using hand methods such as tongs or by power-driven dredges. Some culturists use bags to protect oysters from predators and to ease handling and harvest (fig. 10.1). Bags are made of UV-protected black polyethylene of different configurations, mesh sizes, and mesh shapes. Mesh can be square or diamond-shaped with mesh sizes from 1.5 mm (0.06 inch) to 4.8 mm ³⁄₁₆ inch). The smallest oysters are placed in the smallest mesh bags so that none are lost during handling or growth; the oysters are transferred to larger-mesh bags, concomitant with their growth; larger mesh means more water flow, more food for the oysters, and less probability of clogging the mesh with sediment, algae, and fouling organisms. Some bags are shaped like pillowcases but another popular design, configured with square mesh bags, resembles a suitcase. With the suitcase design, the lid can be opened easily to allow the shellfish farmer to cull or partially harvest the contents of the bag without completely emptying it, thus saving time and labor, an important consideration in intertidal locations where the window of opportunity for working the shellfish farm may be only an hour on either side of low tide.

Aside from bags, several other types of growing equipment have been developed for oysters, thus providing more options for the oyster grower.

Cages have been popular, although they can be heavy and bulky. The Aqua-tray system is an Australian invention that works similar to suitcases in terms of ease of access but has additional advantages of cage culture. Each tray can be divided into several compartments with plastic partitions that prevent the oysters from amassing to one side of the tray with tidal move-ment or rough seas (fig. 10.2). This keeps the oysters more evenly dispersed and assures less competition for food as water flows through the trays. The trays are molded plastic, lightweight, and easily stacked or nested so that layers of trays can be used to maximize the lease area. There are many pos-sible configurations for deploying bags and trays. For example, on his East Dennis oyster farm, John Lovell sometimes secures bags for oyster growing on top of his nested trays. Another type of container for growing oysters is a Seapa bag, a firm, molded, cylindrical bag that can be suspended from a long-line with hooks (fig 10.3). The bags rock back and forth on the lines with the current and tides, which readily and evenly distributes the oysters. The bags are easily removed from the lines and a door-like flap can be opened at one end of the bag to remove oysters for culling or harvest.

No matter which type of bag or tray a grower uses, the gear must be fas-tened securely in some approved manner. In deep-water systems, the bags are often placed in heavy racks or larger trays that can be handled from a boat. On intertidal areas, there are often height restrictions on shellfish growing gear, with 18 inches from the bottom being the most common height allowed by the U.S. Army Corp of Engineers. This allows naviga-tion over shellfish farms during high tides or when the grant is covered with sufficient water. On the intertidal, oyster trays and racks sometimes are se-cured directly to the bottom, but more often they are affixed to racks made from metal rebar (reinforcement metal rods) or PVC pipe. Each grower uses a configuration of racks with adequate spacing to allow ease of access and ease of maintenance and harvest. Some areas allow a floating system for shellfish bags. Instead of being tied to racks, the floating bags are linked together in long rows and secured to the bottom with an anchor. The benefit of a floating system is readily apparent in areas where algae prolif-erate on gear and prevent water flow to the oysters. The floating bags can be turned over periodically so that the sun and heat kills the algae on the upward side of the bag. Among all the types of gear used by shellfish farm-ers in the Northeast, the floating bags are perhaps the most controversial. There are movements in Connecticut and in the towns of Osterville and Cotuit in Massachusetts to ban such gear or to prevent further approvals for expansion of its use. Some nearby homeowners consider them to be

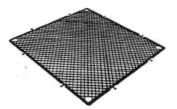

Top Lid (20mm Mesh)

Tray - 4 Partition (12 / 25mm Mesh)
Base and Sides

Tray - 9 Partition (12 / 25mm Mesh)
Base and Sides

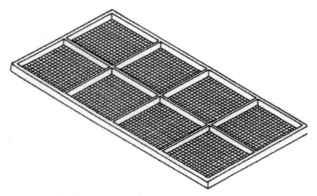

XL Tray - 8 Partition (20 mm Mesh)
Solid Sides

FIG. 10.2 Aquatray system. *Drawings by Peter Dion; courtesy of Coastal Aquaculture Supply*

FIG. 10.3 Seapa bags; field placement in Eastham. *Photo: Barbara Brennessel*

eyesores, but the major reason for opposing this type of oyster-growing system is that of multiuser conflict: The floating bags severely limit access to waters that are also used by boaters, kayakers, swimmers, and those who want to fish or shellfish. The growers who use this system contend that, because of the high algae levels, they will be put out of business if they have to revert to a bottom-growing operation.

Quahog Equipment

With the exception of a unique clam-bag system used by quahog growers in Florida, most quahogs are grown directly on the bottom, with nursery and grow-out in the same bottom plot. The long, rectangular tracts used for quahog cultivation usually are mapped out in advance and of suitable size so that they can be covered with protective netting (see fig. 8.1). The shellfish grower prepares the bottom by raking to remove epifaunal predators and then the small quahogs are scattered evenly onto the plot at an appropriate density, which depends on the size of the seed. In Wellfleet, the 12- by 50- or 12- by 100-foot (3.6- by 15- or 30-meter) quahog bottom plots or intertidal raceways are common, but no matter the location, small quahogs exhibit better survival when the plots are covered with plastic predator netting (some types of netting were developed to cover grape vines to protect the grape crop from being consumed by birds). So that predators can't crawl underneath, the nets must be fastened securely around the quahog bed with materials such as rebar, chain, PVC pipe, or long flexible bags filled with sand (Castagna 2001). To further secure the netting, the surrounding structures may be secured to the bottom with bent rebar "staples." If quahogs are netted, it is important that they are free to move toward the water/sediment interface. If sediment becomes trapped and accumulates below the net, their siphons may not be able to reach through the sediment and they may not be able to breathe. Thus, an entire crop could be lost. In areas that tend to get covered with sediment, growers sometimes uses blocks of styrofoam or another type of spacer to keep the nets floating above the bottom and thus give clams access to the surface. The nets must be checked periodically for holes and any accumulating algae must be removed by brushing it from the net.

Scallop Equipment

Bay scallops also can be grown out in the field if a deep-water site is available. Because bay scallops are mobile bivalves, their culture varies from that of oysters and quahogs. Attempts to farm scallops in the Northeast are modeled after methods that have proved successful in Asia. The most popular gear design is manufactured from mesh over a wire frame, such as the lantern net made in Japan for growth of pearl oysters. These consist of individual round or square growing chambers that are attached vertically to one another and suspended in the water column, usually on long-lines. The chambers have a door or velcro flap to allow easy access (fig. 10.4). Lantern nets also can be used to grow sea scallops.

Coastal Aquacultual Supply reminds growers of the "5-meter rule" developed in Japan, which recommends that the lantern nets be strung below the uppermost 5 meters (15 feet) of water and above the bottom by at least 5 meters (15 feet), as a compromise to limit fouling and produce adequate growth. Currently, bay scallops cannot be cultured in sufficient quantity to make most operations profitable because it has proved difficult to get bay scallops large enough by the winter of their first year with any of the standard aquaculture techniques. Scientists are looking for genetic stocks of fast-growing bay scallops that can be harvested from aquaculture operations at the end of their first season and thus prevent massive winter mortality.

FIG. 10.4 Lantern nets for bay scallop culture. *Photo: Barbara Brennessel*

Surf Clam Equipment

For grow-out, surf clams require deep water and are therefore difficult to farm. Nevertheless, some folks are experimenting with surf clam culture. Surf clams were once abundant in Cape Cod Bay and Provincetown Harbor, but shellfish companies from the entire eastern seaboard would hit the site hard, year after year, with dredging operations. Boatloads of surf clams were destined to be the raw material for commercially produced clam chowders. The Cape Cod towns of Truro and Provincetown enacted a ban on surf clam dredging, so now entrepreneurs like John Baldwin of Seafood Divers Inc. harvest them from 5.5 to 7.6 meters (18 to 25 feet) of water with scuba gear in summer and special dive suits with warm-water inflow during winter. John is a pioneer in the cultivation of surf clams, maintaining a surf clam nursery on a barge to raise seed to planting size on his deep-water "grant" in Provincetown, banking on the probability that small, farmed surf clams may be an alternative for softshell clams in the seafood market and in "steamer" preparations in restaurants.

No matter the method, grow-out is a busy time for shellfish culturists in the Northeast. The rhythm of the tides and the nuances of the weather dictate the grower's work schedule. There is something to do almost every day, so that the term "weekend" has no relevance in this profession.

Chapter 11

Microscopic Troublemakers

During his trip along the coast of Massachusetts in the 1860s that inspired his collection of stories about Cape Cod, Henry David Thoreau found, cooked, and ate a large clam for dinner before staying at the home of a Wellfleet oysterman in the dunes near Newcomb Hollow (Thoreau [1865] 2004). Shortly after eating the clam, he experienced intestinal discomfort that he attributed to his clam repast. Although the exact nature of Thoreau's short-term illness was never diagnosed, the occurrence of shellfish-related diseases in humans is well documented. Eating shellfish, particularly in the days before health and sanitation programs were in place, could be hazardous to one's health. The most common types of shellfish-related illness can be traced to biotoxins, produced by marine phytoplankton, or to microorganisms such as bacteria and viruses. Often, the most-minute organisms can cause the most catastrophic problems. Although many microbial culprits are naturally occurring, others can be traced to contamination of water by human waste.

Harmful Algal Blooms

Phytoplankton are important food for bivalve molluscs, but some species produce toxins that can cause illness in people who consume shellfish. About twenty dinoflagellates of the more than 1,500 known species, as well as some diatoms, have been found to produce toxins that can harm humans. The consumption of shellfish that contain these so-called biotoxins can cause anything from mild intestinal discomfort, such as that experienced by Thoreau, to paralysis and even death. Biotoxin accumulation in shellfish is a global problem, with different algal species contributing to the issue in different parts of the world. The various types of toxins cause ASP (Amnestic Shellfish Poisoning), DSP (Diarrheal Shellfish Poisoning), NSP (Neurologic Shellfish Poisoning), and PSP (Paralytic Shellfish Poisoning), with each malady caused by a different toxin and named for the most pronounced symptom. For example, ASP, which has made its appearance in the Gulf of Mexico, is caused by the biotoxin domoic acid, which may cause disorientation and confusion in addition to nausea, vomiting, and diarrhea. In higher doses or in more sensitive individuals, domoic acid causes more severe symptoms such as headache, seizures, hallucinations, and memory loss. DSP causes diarrhea mediated by the biotoxin okadaic acid. NSP is caused by brevetoxin, which causes gastroenteritis, muscle ache, and dizziness. Unlike other shellfish toxins, brevitoxins, at high concentrations in the near-shore waters, can become airborne and produce allergy-like symptoms in sensitive individuals. PSP results from exposure to saxatoxin, a biotoxin that causes tingling and numbness of the lips, tongue, and extremities, drowsiness, giddiness and/or unsteadiness, vomiting, and diarrhea. Symptoms usually appear within 30 minutes after ingestion. Death is rare but may occur as a result of respiratory arrest.

Events that result in the rapid and prolific growth of specific toxin-producing phytoplankton are technically known as harmful algal blooms (HABs), but are more commonly referred to as red tides. The algae or diatoms that cause HABs are usually naturally occurring and present at low densities. The term red tide stems from observations that massive growth of some species can cause the water to change color, as was probably the basis for the account of water turning red in the story of the plagues that were visited upon Egypt in the biblical story of Moses. There is speculation that the Jewish kosher dietary laws, which approve fish with scales but consider shellfish and crustaceans to be "unclean," stem from occurrences of HABs. Most often, HABs are not detectable to the naked eye and the color of the water is not altered, so the term "red tide" is not very accurate. Other red tides that visibly color the water red or brown are completely harmless to humans. Furthermore, some blooms may be nontoxic to humans but lethal to fish, shorebirds, and marine mammals such as whales and porpoises (Whitaker 2007).

Small, localized blooms occur with some regularity in estuaries and basins that are poorly flushed. For example, Cape Cod locations such as Nauset Marsh and Salt Pond in Eastham and Town Cove in Orleans experience almost annual red-tide events. However, in May of 2005, a particularly extensive HAB occurred from Maine to Massachusetts, the most serious and widespread since 1972. Shellfish beds were closed throughout Maine, New Hampshire, and parts of Massachusetts for weeks. Not only were the shellfish farmers feeling the impact, but tourists changed their plans to visit Cape Cod, causing an economic plague throughout the region. Ripples were felt for months, as restaurants scrambled to get shellfish from the south while many consumers just stayed away from shellfish altogether.

The 2005 HAB was caused by an algal species called *Alexandrium fundyense,* a single-celled dinoflagellate, invisible to the naked eye. A related *Alexandrium* species, *A. catanella,* has been responsible for HAB outbreaks in California. *Alexandrium fundyense* usually is present in the water at low levels. As part of its life cycle, *Alexandrium* can encapsulate into dormant cysts that settle into the sediments, allowing the dinoflagellate to remain in a resting state throughout the winter or even for several years before germinating. Scientists are still uncertain about the exact circumstances that led to the 2005 outbreak, but environmental conditions such as warm surface-water temperatures, lowering of salinity, increases in certain nutrients, and calm seas seem to correlate with HABs due to *Alexandrium.* The winter and spring of 2004–2005 were particularly wet, resulting in a large influx of freshwater into coastal embayments. After germination, the algae usually are carried in a north-to-south direction by currents in the Gulf of Maine, but environmental conditions during spring 2005 also caused it to spread further south, where easterly winds blew it into shore and into shellfish growing areas, rather than out to sea.

When molluscs filter water as they consume phytoplankton, they filter *Alexandrium* and concentrate the PSP-causing toxin, which is still active even when the shellfish are cooked. Contaminated shellfish cannot be identified by taste or appearance. Consuming other seafood species such as lobster, shrimp, and finfish does not pose a risk because these species do not concentrate the toxin in their digestive systems. Scallops from HAB-contaminated waters are safe to eat because the toxin does not accumulate in the scallop muscle and usually only the adductor muscle of the scallop is consumed. If whole scallops were eaten, they would be just as pathogenic as other bivalves taken from *Alexandrium*-contaminated water.

HAB Monitoring

Because of the severe threats to human health that can be caused by red tides that are not visually apparent, all northeastern states have programs in place to monitor for HABs. In Massachusetts, HAB monitoring and resulting shellfish closures are the responsibility of the Department of Marine Fisheries, under Massachusetts General Laws, Chapter 130, Section 74A and 75. From early April through November, shellfish are sampled at weekly intervals from fixed stations throughout the state. When saxatoxin levels reach 50 µg per 100 mg of shellfish tissue, sampling is extended to other sites and is done at more frequent intervals. Shellfish beds are closed when the concentration of toxin reaches 80 µg/mg of tissue of indicator shellfish, usually blue mussels. Mussels have been chosen as the sentinel for toxin concentration because of their tremendous filtration rates, but other shellfish are also tested when the concentration of *Alexandrium* is on the rise. The Department of Marine Fisheries sends out notices to towns, shellfish wardens, environmental police, state agencies, and all personnel who monitor HAB events. After the bloom subsides, monitoring continues with regular frequency until toxin levels subside to below 80 µg/100 mg of tissue. Shellfish beds are not open until toxin levels remain under 80 µg/mg tissue for three consecutive samplings. Different shellfish species clear the toxin at different rates, so it may be possible to harvest one type of bivalve while harvesting others is still prohibited. Toxin levels in surf clams tend to remain high even when oysters and quahogs are safe to eat. Maine has established a Red Tide and Shellfish Sanitation Hotline to provide an up-to-date inventory of shellfish bed closures due to HABs as well as to bacterial contamination.

Some scientists have proposed that monitoring be extended beyond measurement of toxin levels and also include sampling for *Alexandrium* cells. State agencies have worked with scientists such as Dr. Donald M. Anderson at Woods Hole Oceanographic Institute (WHOI) to look for *Alexandrium* cells after increased freshwater infiltration caused by heavy rain. For example, by May 10, 2005, *Alexandrium*—normally present at 200 cells per liter of seawater—had increased to 40,000 cells per liter. Routine water monitoring for *Alexandrium* cells may be able to detect changes in the concentration of algae, and thus constitute an early alert for the possibility of red tide, even before toxin levels in shellfish change dramatically. Scientists are developing systems, methodologies, and models to predict outbreaks of HABs based on environmental conditions and densities of dinoflagellates in order to forecast possible closures.

Once a red tide hits, the shellfish farmer must stand by and wait, a response that can result in considerable loss of income. In 2005, over 1.3 million acres (over 77 percent) of marine coastal waters in 42 towns from Maine to Massachusetts were closed to shellfishing. The red tide shut down the operations of about 2,000 commercial shellfishermen and over 250 shellfish aquaculture operations. Soon after the 2005 HAB made its appearance, a State of Emergency was declared by the governor, and Massachusetts senators and congressmen sought relief from the Federal Emergency Management Agency (FEMA). A similar State of Emergency was declared in Maine. A Red Tide Relief Fund subsequently was put into place to assist some of the shellfish farmers who suffered considerable losses during the 2005 outbreak, with about $2 million given to each of the states seeking relief for the eight- to ten-week period when shellfishermen could not harvest. Some predicted that they would lose almost half of their annual income. In 2007, criteria were put into place so that relief funds could be allocated in an equitable manner and shellfishermen who were affected by the 2005 HAB were invited to informational meetings and given instructions for submitting applications.

Other types of harmful algal blooms do not involve toxins that cause human illness but may involve factors that affect shellfish life and growth. Macroalgal blooms, resulting from increased nutrient input from land-based sources, result in growth of seaweeds that indirectly affect shellfish by altering habitat and shading water so that eelgrass and microscopic phytoplankton cannot grow. In addition, as the macroalgae die, they sink to the bottom, where bacteria busily work on their decomposition. In doing so, the bacteria use up the oxygen in the water, thus creating anoxic conditions that may result in death of fish and shellfish that are denied a source of oxygen.

Another algal bloom, caused by brown algae that are not toxic to humans, drove the nail into the coffin of the Peconic Bay shellfish industry in the 1980s. A very small, unicellular algal species, *Aureococcus anophagefferens,* imparted a brownish tinge to waters on Long Island, New York, Narragansett Bay, Rhode Island, and as far south as Barnegat Bay, New Jersey. My father-in-law, who loved to swim in the warm waters of Peconic Bay, boycotted his town beach for many years during this episode, maintaining that the water felt "slimy." It was not only unusual that the brown tide occurred, it was also unusual that it recurred for many years. The reasons for the occurrence and self-perpetuation of this algal bloom are unresolved but may include increased nutrient loads, resuspension of cells from sediments, and lack of grazing by zooplankton. Whatever the cause, the Peconic Bay brown tides have had a severe impact on shellfish resources and just about decimated the oyster, quahog, and scallop fisheries in the area. In addition, the resulting lack of light penetration caused a massive die-off of eelgrass beds, important nursery habitat for bay scallops and many other species.

Scientists have proposed a number of reasons for the global increase in algal blooms and HABs. Aside from natural mechanisms such as currents and tides that disperse algae, environmental events such as long-term climate change and global warming also may be contributing factors. Direct human activities that enhance nutrient enrichment have been implicated, such as finfish aquaculture and run-off from point and nonpoint land sources, transport and dispersal by ship ballast water, and perhaps even shellfish seeding or transplantation. The improved methods for detecting HABs also may contribute to perceived increases in their occurrence.

Infectious Diseases

Aside from possible contamination with biotoxins, eating shellfish, particularly raw shellfish, may have other hidden dangers for the consumer. Microscopic bacteria and even smaller viruses, present in the water, may be concentrated in shellfish tissues and passed along to unwary seafood connoisseurs. These disease-causing microbes may be present naturally in coastal water, but more often they are introduced by humans and their activities. The problem usually can be traced to coastal waters contaminated with untreated or partially treated sewage. Other microbial problems may occur after harvest when the shellfish are handled, stored, or transported.

The most serious and debilitating in the spectrum of infections is cholera, caused by the comma-shaped bacterium *Vibrio cholerae* o1 serogroup, which has been virtually eradicated in the United States. However, other *Vibrio* species that inhabit coastal ecosystems have the potential to cause gastroenteritis with varying levels of severity. Although *Vibrio* are naturally occurring in marine environments, their distribution and concentration vary with salinity, temperature, and other environmental conditions. *Vibrio* seem to grow best in warmer, brackish waters, and most U.S. cases have been associated with oysters from the Gulf of Mexico (Rippey 1994).

Most other shellfish-associated infectious diseases can be traced back to the development of sewage systems, beginning in the late 1800s, which delivered concentrated untreated human waste to local water bodies, and to the construction of impervious roadways that allowed heavy rains to flush microbes from land-based sources into nearby waters. Thousands of cases and hundreds of deaths were associated with typhoid epidemics caused by *Salmonella typhi,* a pathogenic bacterium that can be transmitted via the fecal-oral route. Some of these cases were traced to contaminated shellfish, providing federal and state agencies with the impetus for the formation of shellfish sanitation guidelines and regulations for growing, handling, and storage of shellfish under conditions that prevent contamination and subsequent human health issues. Although typhoid outbreaks have not occurred in the United States for half a century, other microbial diseases, usually causing symptoms of gastroenteritis (nausea, vomiting, and diarrhea), are responsible for sporadic outbreaks.

Polio is no longer a problem, due to the nationwide, mandatory vaccination program. However, before the vaccine was available, some cases of polio virus infection may have been transmitted by shellfish. Today, the most problematic of the microbes belong to the enteric groups such as the fecal coliforms and viral pathogens such as the Norwalk and Hepatitis A viruses. After consuming contaminated shellfish, the resulting gastroenteritis usually manifests itself in 24 to 48 hours and lasts a day or two. *Escherichia coli* (*E. coli*) is the best-known member of the coliform bacteria group, but related organisms such as *Salmonella, Shigella,* and *Campylobacter* also have been implicated. The coliforms in sewage are destroyed if the sewage effluent is treated properly, usually with chlorine, before it is discharged. Untreated sewage, which may leech out of septic systems and make its way to the water, is referred to as nonpoint-source pollution.

Even though most *E. coli* strains do not cause disease and they may originate in the environment from a variety of warm-blooded animals including waterfowl, this species of bacteria is used routinely as an indicator of fecal contamination because it is harbored in all humans. Detection of its presence by health officials not only closes recreational swimming areas but can shut down shellfish beds. Heavy rain events that result in increased freshwater runoff from land-based sources such as impervious roadways cause an increase in the concentration of *E. coli* in coastal waters. Scientists and coastal managers have debated the use of *E. coli* as an indicator of water quality. For example, *E. coli* might be present within acceptable limits but this tells us nothing about viral concentrations. Conversely, *E. coli* levels may be high enough to restrict shellfishing but the levels of true pathogens may not be significant. Nonetheless, fecal coliform tests that target *E. coli* are simple and inexpensive and have become the method of choice for certifying shellfish areas.

Pathogenic bacteria and viruses may be introduced to shellfish postharvest, for example, if shellfish are stored or shipped with raw meat, poultry, or finfish under conditions that allow contamination or if shellfish are stored in tanks containing contaminated water. Furthermore, if shellfish are not stored at proper temperatures, whether in air or in water, these microbes could flourish.

Protecting the Public

Some of the most productive shellfish areas are located in contaminated waters. New York State has a long-time industry that revolves around moving shellfish from contaminated areas around New York City to clean grow-out areas in Peconic Bay (see chapter 9). Other states have developed similar shellfish-relay programs. Some shellfish harvesters and culturists have access to facilities where shellfish can be taken to clean themselves out. These shellfish "spas" are called depuration plants. As an example, softshell clams are taken from specially designated beds around Boston Harbor that are classified as restricted, and transported by harvesters to a depuration plant. The plant, constructed in 1928 on Plum Island in Newburyport, Massachusetts, is the oldest continually operating facility in the country. About 560 bushels of clams are cleansed each week with sterilized seawater in a process that takes a minimum of three days. If the clams contain more than 1,600 coliforms per 100 grams of meat, they are rejected by the facility and confiscated, due to the belief that these bacteria-loaded clams will not purify adequately during the depuration process. The facility has an excellent track record; no record of illness can be traced back to clams depurated in Newburyport.

Matthiessen (1992) points out the pros and cons of depuration. On the plus side, clams that are unsuitable for consumption are converted to a utilizable resource and consumers are assured a safe product. On the minus side, it can be expensive for the harvester to send clams to these facilities and the process does not address the root of the issue: unacceptable water quality. As an added problem, depuration may clear fecal coliforms but fail to remove viruses lodged within shellfish tissues. Furthermore, stressed or damaged clams may not be able to filter the sterile water and thus remain a health hazard.

The National Shellfish Sanitation Program (NSSP), created in 1925, protects public health by overseeing shellfish issues related to contamination of water in growing areas by biotoxins or human waste and determining whether shellfish are safe for human consumption. NSSP, using guidelines promoted by the U.S. Food and Drug Administration (FDA) has developed uniform standards that are used by various states to monitor water quality and to harvest, process, and distribute shellfish. The standards cover sanitation issues in growing areas, including land-based and open-water aquaculture sites. In general, areas are approved for shellfish aquaculture after 15 samples have been taken under different tidal and environmental

regimes and found to contain a median number of fewer than 70 total coliform bacteria per 100 ml and not more than 10 percent of the samples
have a count in excess of 230 coliforms per 100 ml, or the median number
of fecal coliforms (*E. coli* as indicator) is less than 14 per 100/ml and not
more that 10 percent of samples exceed 43 fecal coliforms/ml (Massachusetts Office of Coastal Zone Management 2007).

NSSP also has protocols for laboratory procedures and testing, as well
as for the certification of shellfish shippers. Participation in NSSP is not
compulsory; however, states that do not adhere to the guidelines are prohibited from selling shellfish across state lines. In Massachusetts, the Department of Marine Fisheries, working under NSSP protocols, has the role
of testing water at 2,320 stations in 294 shellfish-growing areas at least five
times during harvesting season (Hickey 2002). In Connecticut, shellfish
sanitation is overseen by the Department of Agriculture, Bureau of Aquaculture. As in other states, growing areas are classified in several categories
based on the number of coliform bacteria: approved, conditionally approved, restricted, conditionally restricted, and prohibited. This agency also
oversees Connecticut's seed oystering program, which permits the harvest
of seed in conditional, restricted, and even prohibited areas of public bottomlands. The seed is then sold to licensed individuals or companies that
relay or transplant them to approved growing areas (Getchis, Williams, and
May 2006). In contrast, Massachusetts does not permit shellfish relay from
uncertified waters by private companies (Matthiessen 1992). In New York,
the Department of Environmental Conservation shares responsibility
with the Department of Health and Agriculture and Marketing for monitoring the health of shellfish-growing areas.

In all states, local town and county shellfish officers are responsible for
enforcing the harvesting regulations and preventing harvest from conditional and prohibited areas unless an approved relay is in progress. The federal government is involved in assuring the safety of shellfish, whether it is
wild or farmed. The Interstate Shellfish Sanitation Commission (ISSC),
formed in 1982, is an important partner of the U.S. Food and Drug Administration and was formed to foster cooperation among state and federal
agencies, the shellfish industry, and scientists, and to provide a forum to re-
solve issues related to shellfish sanitation.

The Taunton River Estuary in Massachusetts was a major oyster-
producing area, but the flesh of oysters turned green, presumably due to
pollutants from an upriver copper works (Matthiessen, 1992). This inci-
dent shows that biotoxins and microbes are not the only means that can

make shellfish unfit for human consumption. Pollution, caused by chemicals, oil spills, and industrial waste, also has the potential to be concentrated in shellfish tissues. The shellfish regulatory agencies routinely test for a variety of known chemical pollutants that may make their way into those oysters at the raw bar or into your clam chowder. With all the current safeguards in place, shellfish have become a safe and savory food choice, whether they are sold in a seafood store or the all-purpose supermarket or if they are on the restaurant menu.

Chapter 12

Dangers Abound during Summer on the Flats: Common Bivalve Predators

My neighbor Russ is semi-retired. When he left his former trade, he started to harvest oysters commercially, just to keep busy and supplement his income. After picking wild oysters for a few years, he decided to apply for a "grant" and give oyster farming a try. If you ask Russ if he is working hard, he'll tell you that he's hardly working. But after one day out with Russ on his "grant," I realized how much time and back-breaking effort he is devoting to his oyster farm to make it pay off.

Aquaculturists like Russ live by the tides. The ebb and flow of tidal movement provide the rhythm, schedule, and structure for each day. At low tide, the shellfish farm is often more accessible to the grower. During the long days of summer, it is sometimes possible to work two tides per day. Trays and bags are tended and there is plenty of other work that must be done. Oysters that have grown and are becoming crowded are separated into roomier accommodations. Often, more bags and trays must be deployed to provide space for the oysters to grow and to give them access to the phytoplankton that sweep across the grant with the currents and tides. Oysters are culled. Those that are stuck together are gently chipped apart; oysters with misshapen shells are separated from their counterparts that are growing with the desired symmetry. Like Russ, most commercial growers in the Northeast are rearing "boutique" oysters for the half-shell market. Not only do they have to taste good, they also have to look good. Beauty contest criteria include a nice deep cup, a clean, fan-shaped shell, and of course, the ideal size: large enough for legal harvest but not too big and chewy. As the farmer tends his or her crop, some oysters may be ready for market and can be harvested, usually into bushel baskets. In deep-water oyster farms, such as those in Long Island Sound, boats with hydraulic equipment pull up bottom trays and gear. Culling and harvesting are performed on board. Boats are also required for scallop and mussel farmers to tend their crops.

At low tide, the quahog farmer has the opportunity to check protective netting for holes and make appropriate repairs. Nets that are covered with seaweed and other organisms are cleaned or replaced. Quahogs are checked for growth, but unlike oysters, which need constant attention, the quahogs need little care until they are market size and ready for harvest.

Despite the constant attention and oversight the aquaculturist provides, at times, the fate of the harvest is governed by factors beyond his or her control. The overall health and biological condition of bivalves can be affected by other living organisms that share the aquatic environment. Microscopic invaders such as viruses, bacteria, and single-celled protists are capable of causing infection, disease, and even death in natural as well as cultivated shellfish populations, thus decreasing the commercial value of the shellfish crop. These shellfish enemies commonly are referred to as parasites, because they reside and multiply within the tissues of the bivalve, or pathogens, when their colonization results in outright disease in their bivalve host. Some of the symptoms of these diseases are impairment of growth, reduced condition of the meat, inhibition of reproduction, and increased mortality. Some microbial diseases and pathogens are unique to hatchery operations. Despite the nature of the problem, some of the smallest organisms can do the most damage.

Bivalve populations are also prey for various predators that feed on them in specialized ways. Predators fall into several biological categories, from small worms to large vertebrates such as humans. The bivalves themselves may even eat their young. This cannibalistic occurrence results when a large bivalve is filter feeding. In vacuuming up phytoplankton, it also may take in gametes and larvae of its own as well as other bivalve species.

Although bivalve molluscs of commercial importance have hard shells that make them impervious to most types of penetration and predation, several types of predator are part of the coastal ecosystem and have persisted over time along with their molluscan prey. However, localized imbalance in their numbers may cause a dent in commercial shellfishing operations and these predators are categorized as enemies to mollusc growers.

But all things considered, these predators are just trying to make a living; they like to eat clams and oysters as much as we do. Predators feed heavily on the very youngest molluscs. Even though they inflict a high mortality on spat and juveniles, the fecundity of adult molluscs has evolved to compensate for the high losses. However, mollusc evolution has not kept pace with the ability of humans to take their toll on a bivalve population. We take them at their prime, often in great numbers and often after they have reached sexual maturity. Humans are, by far, the most deadly enemy.

Predators of bivalves are difficult to control and can have a direct effect on the livelihood of growers. They come in all sizes and shapes and have interesting strategies and mechanisms to penetrate the shell.

Worms

The oyster flatworm, *Stylochus ellipticus,* is an elusive predator. It is also known as the oyster leech or oyster wafer. It is oval-shaped, very thin, with a wrinkled outer edge and can achieve a length of 25 mm (1 inch). It inhabits the bottom and is often found under rocks but it can also slip in between the oyster's valves when it is feeding and consume the oyster from within. Spat and juvenile oysters are particularly vulnerable. Although its name suggests that it eats only oysters, barnacles are a favorite food. In 1961, the oyster flatworm was implicated in a mass mortality of oyster spat in an oyster pond on the island of Martha's Vineyard. *S. ellipticus* was the only predator found during the mortality event and it is believed that the dense cultivation conditions may have been a contributing factor. There is only one way known to get rid of these predators in aquaculture operations and that is to immerse bags or trays containing oysters into freshwater or very salty water. The oyster flatworm is sensitive to these conditions and will be destroyed. However, re-infestation can occur (White and Wilson 1996; Flimlin and Beal 1993).

The milky ribbon worm, *Cerebratulus lacteus,* inhabits the intertidal zone. It buries on the bottom and preys on various species of clams, especially softshell clams and razor clams, but quahogs may also be at risk. It has a proboscis with teeth that it injects into the clam and it eats the meat within the shell. All sizes of clam are at risk. These worms can get very large, some achieving 0.9 to 1.2 meters (3 to 4 feet) in length.

A type of worm in the genus *Polydora* can have an impact on oysters. They are not true predators since they don't eat oysters and therefore they are sometimes categorized as pests. However, their growth causes the appearance of blisters at the base of the adductor muscle. As a result, oysters have difficulty closing their shells and thus succumb to other stressors. *Polydora* are inhibited at high salinity (Flimlin and Beal 1993).

Carniverous Gastropods (Snails)

The oyster drill or screw borer, *Urosalpinx cinera,* is found along the east and west coasts of the United States and Canada (fig. 12.1). It has a thick,

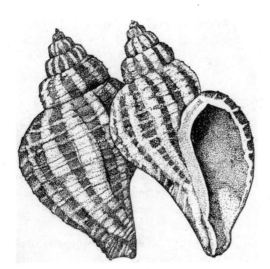

FIG. 12.1 Oyster drills. *Drawing: Marisa Picariello*

spiral shell and can grow to 25 mm (1 inch) in length. Another similar drill, the thick-lipped drill (*Eupleura caudata*) has a range that largely overlaps that of *Urosalpinx,* but it is limited to areas south of Cape Cod. Although this creature is called the oyster drill, it will take advantage of the nutritional value of many types of bivalves as well as barnacles. Oyster drills are even known to be cannibalistic and put each other on the menu. They methodically bore a small hole through the shell of their prey with a combination of chemical and mechanical forces. They produce and secrete chemical compounds that soften the shell material, while using their radula (a mouth part with sharp teeth) as a mechanical drill. After making a hole, they insert their proboscis and consume the oyster. Oyster drills can be responsible for considerable mortality of spat and small oysters. They can bore into larger oysters but it requires more effort on their part.

Oyster drills require salinity above 15 ppt (parts per thousand) for optimal function and reproduction but they tolerate short-term fluctuations in salinity. Temperature is also an important factor in their activity; they are most active between 10 and 30°C (50 and 86°F). Above or below this range, they tend to burrow in the mud and remain inactive.

It is difficult to control oyster drills because they reproduce in great numbers and young drills can crawl away from their home and find other areas with small oysters on which they can feed. In addition, they can be transferred along with their oyster hosts during planting. Baited oyster drill traps have met with mixed success. The invention of powerful hydraulic suction dredges provided large-scale oyster drill elimination opportunities for oyster companies in the early 1950s. To prepare bottoms before shell material was deposited to catch spat, a suction dredge aboard a large oyster boat would vacuum the bottom. As drills were collected, they were treated with very hot gases (a bit over 1000°C or up to 2000°F) and then dumped overboard (Galpin 1989). A chemical called Polystream, made by the Hooker Chemical Corporation in Niagra, New York, was used in the mid-1960s. This pesticide had variable toxicity and fortunately for other marine organisms was banned after a few years of field trials (Galpin, 1989). In 1957, John Glude observed that drills will not cross a copper barrier. He devised experiments using thin copper sheeting, wire, and screens, which he put into tanks of seawater. Even when bait such as oyster spat were used, drills would not cross copper barriers. In the closed system in the laboratory, the drills eventually succumbed to the toxic effects of the copper as it leeched from the copper-containing barriers (Glude, 1957). Despite the observed potential of copper as a barrier to oyster drills, it has not been utilized widely to design gear for the purpose of drill control. Some growers have experimented with copper barriers on their own equipment and have met with mixed success. As Glude concluded in his study, before such deterrents are employed, it would be important to determine the extent of copper ionization and its effect on oysters and other species.

For aquaculture applications, the most effective method to prevent high mortality by oyster drills is to rake the area under culture to remove all drills and use netting and small-mesh trays or cages to protect young oysters. Flimlin and Beal (1993) also report that aquaculture gear can be dipped in saturated salt solution and air dried. Although this procedure will inhibit drills, it may also put stress on the bivalves being cultured.

Moon Snails

Moon snail shells are round, smooth, and light tan or grey in color (fig. 12.2). The northern moon snail (*Euspira heros*) can reach 7.6 cm (3 inches) in diameter and its shells are commonly seen on beaches in the northeastern United States and into Canada. A complete, unbroken shell is a beachcomber's prize. The Atlantic moon snail (*Neverita duplicatus*), commonly called the shark's eye, is somewhat smaller than the northern moon snail. It ranges from Cape Cod south to Florida. Moon snails are shellfish predators that glide along the bottom on a large, mucus-covered foot. At low tide, moon snails often can be found under the mud. When the moon snail

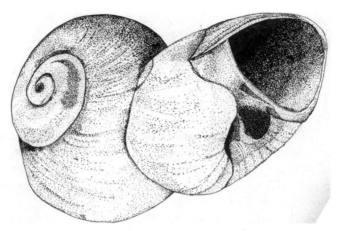

FIG. 12.2 Moon snails. *After Abbott 1968. Drawing: Marisa Picariello*

encounters a clam, oyster, or even another moon snail, it will envelop it with its fleshy foot. The moon snail then uses its radula to drill a hole in the shell of its prey. The drilling is aided by chemical secretions that soften the shell. The drilling process is slow and methodical and results in a beveled hole, usually near the hinge of the bivalve. The prey is eaten through the newly formed hole.

The moon snail discharges its eggs along with a volume of mucus that it molds with its foot. The resulting reproductive structure takes the shape of a round collar. Sand adheres to the collar and helps it to retain its circular shape. The collar will never form a complete circle because the snail starts the egg laying and terminates it without joining the ends. The snails pass through several larval stages before they develop into tiny moon snails.

Whelks

Two species of whelk, or conch as they are commonly called, the knobbed whelk (*Busycon carica*) and the channeled whelk (*Busycon canaliculatum*), are not dissuaded by the size or strength of the adult bivalve's shell (fig. 12.3). They are not even discouraged if the bivalve is buried in the sediment. They grab the clam or oyster with their large, powerful, muscular foot, then use their foot as a vice and their own shell as a wedge to access the inner contents of the bivalve's shell. Once the shell is damaged, the powerful foot is used to pry the valves open enough so that the whelk can

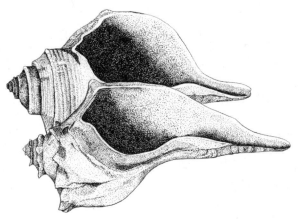

FIG. 12.3 Whelks: *Busycon canaliculata* on top and *Busycon carica* on bottom. *Drawing: Marisa Picariello*

insert its proboscis (mouth parts) and suck out the flesh. Chips along the outer edge of a live clam or oyster shell are fieldmarks of an unsuccessful predatory attempt by a whelk. The reproductive structure of the whelk is a very long, distinctive, ribbon-like structure with stacked papery discs, each containing hundreds of progeny that are miniature versions of the adults.

A simple system for controlling whelk predators is to promote their culinary qualities. Whelks are edible and considered a delicacy in some cultures. If more people appreciated the interesting flavor and texture of whelks, another type of shellfish market could be created and the number of whelks preying on clams and oysters might decrease. Unfortunately, the preferred bait for whelk is the horseshoe crab (*Limulus polyphemus*), so environmentalists are concerned that the "conch" fishery has the potential to deplete the population of horseshoe crabs, which are important members of coastal ecosytems.

Crustaceans

Crabs are among the most damaging predators of cultured shellfish. They also take their toll on natural clam and oyster populations. They can crush small shellfish with their powerful claws and chip away at the larger ones along the edges of their shells and then the crab will use its claws or mouth parts to extract the meat. In the Northeast, many species of crab prey on shellfish, including the large blue crab (*Callinectes sapidus*), which can measure 22.5 cm (9 inches) from tip to tip. The blue crab is found along the south coast of Cape Cod and along the coasts of Rhode Island, Connecticut, and New York, and points south. In recent years, it has made its appearance further north, in Cape Cod Bay. The green crab, *Carcinus maenas,* was introduced from Europe in the 1800s. Although it is smaller than the blue crab (8 cm or 3.75 inches), it can have a considerable impact on shellfish, especially quahogs and softshell clams. Other crab species that prey upon bivalves are the aggressive lady crab, sometimes called the calico crab (*Ovalipes ocellatus*); the spider crab (*Libinia dubia* and *Lubinia emarginata*), which prefers small bivalves; juvenile rock crabs (*Cancer irroratus*); various species of small mud crabs in the genera *Rhithropanopeus, Panopeus, Dyspanopeus,* and *Neopanopeus;* hermit crabs (*Pagurus longicarpus*); and the large and familiar horseshoe crab (*Limulus polyphemus*), which is related more closely to spiders than to crabs.

Pea crabs in the genus *Pinnotheres* are tiny symbionts of shellfish. They are typically about 1.3 cm (½ inch) in length. Some reside within oysters, others prefer mussels or scallops. The exact relationship of the pea crab with its molluscan host is not known. For most of their lives, they remain within the bivalve and feed on plankton and particulate material collected on the mollusc's gills. The female pea crab matures within the oyster. The smaller, free-living male will enter an oyster containing a female. After copulation, the male will resume his existence outside of the confines of the oyster. The female remains inside the oyster to produce larvae, which are released for a free-swimming stage. There is evidence that the constant feeding activities of the pea crab can prevent the oyster from obtaining sufficient nutrients and also may damage the mollusc's gills (White and Wilson 1996).

Starfish

Starfish (also called sea stars) such as the common star (*Asterias forbesi*) and the northern sea star (*Asterias vulgaris*) prey on barnacles and bivalves (fig. 12.4). Historically, this predator was a particular problem for the shellfish industry in the Northeast. Starfish use the power of their suction feet to open the bivalves by tiring their adductor muscle. Once the muscle becomes fatigued, the bivalve, cannot keep its shell closed. With a portal of entry, the starfish will evert its stomach (push it out of its body) and insert it into the bivalve where it will digest it in the shell. Some biologists speculate that the starfish secretes an anesthetic that numbs the bivalve and prevents it from completely closing its shell. This is an intriguing hypothesis, but I have not been able to find scientific studies that demonstrate this effect.

FIG. 12.4 The common starfish (*Asterias forbesi*). *Drawing: Marisa Picariello*

Starfish can detect quahogs buried under the mud and quahogs and scallops can sense the presence of a starfish. The quahog will respond to the presence of a starfish by reducing its pumping rate and burrowing deeper in the sediment (MacKenzie et al. 2002), while scallops will use evasive behavior and swim away.

Starfish were once abundant in the Northeast. *Asterias forbesi* is responsible for the vast destruction of oyster beds in Long Island Sound in the early 1900s. During the 1950s through 1970s, aggressive measures were taken to remove starfish from Long Island Sound near the Connecticut coast and from Great Sound Bay off Long Island (MacKenzie and Pikanowski 1999). Starfish have an amazing ability to regenerate. A single limb of a starfish may give rise to an entire animal in a year or so. It does no good to remove starfish from the vicinity of a shellfish bed, chop them up, and throw them back in the water; this will only create more starfish predators. In areas where starfish were abundant, there was some experimentation with the application of lime during slack tide. This measure was very costly and not entirely effective. The only tool that has been found to be valuable in starfish control is the "star mop," which has been in use since the 1800s. This effective device consists of a 3-meter (10-foot) iron bar to which heavy rope or yarn mops are attached at intervals to accommodate 10 to 15 mops. The mops extend about four feet and trail along the bottom. As the mops are dragged by boat, starfish became enmeshed in the yarn. During the massive starfish control programs, the starfish-filled mops were brought on board and placed into vats of steamy water. This procedure served two purposes: It killed the starfish and also caused them to disentangle from the mops (Churchill 1920).

Due to starfish eradication efforts, the cyclical nature of starfish population explosions, and other poorly understood factors, starfish have not been abundant in the Northeast since the 1970s. In the New York–New Jersey area of Raritan Bay, starfish abundance declined after 1992 for unknown reasons. Not surprisingly, the decline in starfish correlated with an increase in bivalves, specifically quahogs (MacKenzie and Pikanowski 1999). Since the mid-1980s, starfish have virtually disappeared from Connecticut oyster beds in Long Island Sound. Although no clear explanation has been found for the decline in starfish in the Northeast, this phenomenon certainly has benefited the shellfish industry.

Vertebrate Predators

A variety of vertebrates represent some danger to bivalve molluscs. Cownose rays (*Rhinoptera bonasus*) prey on bivalves, especially quahogs in southern New England and points south. By beating its wings near the soft or sandy bottom bottom, the ray can stir up the sediment and expose bivalves, which it then consumes after crushing its prey with its teeth.

Fish are stopped by the calcareous boundary of the molluscan shell but are not completely discouraged from preying on bivalves. Their strategy is to take a meal by nipping at the exposed siphons. This damages the siphon but does not necessarily kill the bivalve. Siphon-cropping fish include flounder, drum, tautog, and northern puffer.

Avian predators include eider ducks, which stamp their webbed feet on the tidal flats to detect small shellfish, and oyster catchers, which use their sharp beaks to break the shells. Gulls have learned to gain access to shellfish meat by picking up their prey, ascending, and then dropping their catch onto a hard surface such as a roadway or parking lot to break open the shell.

Wild and cultured shellfish that have survived lurking predators in the coastal environment have not completely escaped from the gauntlet of agents that can cause their demise. Safety is a relative condition and there are additional dangers that are both invisible and unpredictable. The next chapter explores these internal menaces.

Chapter 13

Parasites and Pathogens

Shellfish Infections and Diseases

Many bivalve shellfish do survive predation and as they grow larger and produce their strong shells, they become more resistant to the attacks of predators and thus may live out their normal life span. However, organisms that work from the inside out can take down even the older, stronger bivalves. These microscopic infectious agents sometimes can reside within the bivalve without causing overt problems. After all, their strategy may not necessarily involve killing their host, because if they do, they must move on to another host or else perish. In nature, a balance is often struck between parasite and host. It is only when the balance is tipped against the host that disease ensues. If a host dies, the parasite usually has a method to infect another healthy host and continue its own existence.

Most parasites exhibit some specificity for the hosts they infect. Oyster parasites do not appear to infect quahogs and are often specific with respect to the type or species of oyster they infect. When parasitic organisms are routinely present in a small percentage of a population, the condition is described as enzootic. In contrast, when a sudden, large outbreak occurs, it becomes an animal epidemic, technically known as an epizootic.

Oyster Parasites and Pathogens

MSX

MSX is a disease of oysters caused by the protozoan *Haplosporidium nelsoni*, named after Thurlow Nelson. The disease was first detected in Delaware Bay in 1957. By the mid 1960s, it was seen in North Carolina, Virginia, Maryland, Delaware, and New Jersey. It progressed to Wellfleet Harbor, in Massachusetts, in 1967 and by the mid-1980s, its range broadened from Florida to Maine (Haskins and Andrews 1988; Ford and Tripp 1996). The most likely route of entry of *H. nelsoni* to the East Coast was with the introduction of the Pacific oyster, *Crassostrea gigas,* from California or Asia. In 2002, MSX spread farther north and made its appearance across the U.S. border in the Bras d'Or Lakes, near Cape Breton in Nova Scotia (Bower 2006a). Infection of adult oysters causes their meat to become thin and watery. Warmer water temperatures and high salinity promote the spread of the disease (table 13.1). Although not harmful to humans who eat the oysters, the infection causes a poor condition of the oyster and a decline in its commercial value. Eventually, the oysters can die as a result of the disease. In the Northeast, is most prevalent from July to September.

Several major outbreaks of MSX caused mortality of oysters throughout the 1980s. The causative agent of MSX was a mystery (hence the name MSX: multinucleated sphere with unknown affinity), until it was identified as a protozoan. In each animal, initial MSX infection is seen in the gills but the organism can spread to other tissues. *H. nelsoni* can be seen in oyster tissues as a multinucleated plasmodium (a large cell with many nuclei; one of the stages in the life cycle of certain unicellular organisms) of 5 to 100 μm in diameter. Detection of the organism in tissue and blood smears is the most common method of diagnosis.

MSX is also capable of forming spores (one-celled reproductive structures that potentially can give rise to new individuals), but scientists have not yet elucidated its complete life cycle nor how the disease is transmitted between oysters. There is some speculation that an intermediate host may be involved because the rate of spread of MSX does not always correlate with the density of oysters and the disease can spread rapidly over long distances (Ford and Tripp 1996). In addition, it has not been possible to transmit the disease to oysters in the laboratory by exposing them to cultures of *H. nelsoni.*

Spat do not succumb to MSX, most likely because they filter less water, and hence fewer disease-causing organisms, than do adult oysters. The Pacific oyster, *C. gigas,* appears to be resistant, but scientists do not know the biological basis for resistance. Evolution has been operating in parallel with laboratory breeding programs to produce MSX-resistant oysters. Scientists and growers utilized the survivors of epizootics for selective breeding programs in attempts to produce lines of highly disease-resistant organisms that can be used as broodstock in hatchery operations. This has resulted in several genetic strains of MSX-resistant oysters. Natural selection has resulted in high MSX-resistance in survivors of two separate epizootics in natural oyster populations in Delaware Bay. Oysters that survived MSX for several years in Delaware Bay in the late 1950s were used to

Table 13.1.

Comparison between Major Oyster Diseases of the Northeast

	MSX	*Dermo*
Causative agent	*Haplosporidium nelsoni* Proto-zoan Phylum: Acetospora	*Perkinsus marinus* Protozoan Phylum: Apicomplexa
Portal of entry	Initial infection in gill and palp epithelium	Digestive tract: gut epithelium
Optimal water temperature for spread of infection	5–20°C (41–68°F)	25–30°C (77–86°F)
Optimal salinity for spread of infection	15 ppt (MSX cannot survive below 10 ppt)	Above 15 ppt
Mode of transmission	Unknown	Oyster to oyster via motile spores; also may be spread indirectly by mobile gastropods and other scavengers that feed on diseased oysters
Seasonal nature	Infections seen in late June, most prevalent from July through September	Most prevalent in September and October; spring mortality from oysters infected in previous growing season
Control measures	Use of low-salinity growing areas	Use of low-salinity growing areas

Source: Information for the table was summarized from the following sources: Ford and Tripp 1996; Ewart and Ford 1993.

produce genetic lines of MSX-resistant oysters by Dr. Harold Haskin at Rutgers University. Scientists at the Haskins Lab continue to develop MSX-resistant oysters and improve the strains to include other qualities such as increased growth rates and resistance to both MSX and Dermo disease. Other MSX-resistant lines have been developed at the Virginia Institute of Marine Science. However, resistance is just a relative situation; resistant oysters can still become infected with *H. nelsoni,* but they will tolerate infection better than susceptible strains and can achieve market size before MSX has a major physiological impact (Ford and Tripp 1996). Stress due to shell damage, algal blooms, and decreases in the phytoplankton food supply can contribute to susceptibility to MSX.

No single measure can prevent the appearance of MSX at a shellfish farm. However, shellfish farmers can take some steps to reduce the problem of

MSX. If they can afford to purchase seed, they may consider purchase of resistant strains from hatcheries. If they are using spat from the local growing area, MSX will be less of an issue if they are farming in lower-salinity areas. Although infected oysters should be removed from growing areas, removal alone cannot halt the progress of the disease.

Dermo

Similar to MSX, Dermo is caused by a protozoan organism but, in the case of Dermo, the culprit is *Perkinsus marinus.* The causative agent originally was thought to be a fungus and initially was named *Dermocystidium,* hence the name Dermo. The disease, sometimes called Perkinosis, was introduced accidentally into Pearl Harbor, Hawaii (Bower 2006b). It was first observed in the Gulf of Mexico in the late 1940s and gradually spread eastward along the Gulf and northward along the East Coast. Its rapid spread was assisted by environmental parameters such as drought and warm temperatures in the late 1980s and early 1990s, as well as by unintentional transport of Dermo-infected oysters (Ford and Tripp 1996). Consideration of the patterns of distribution and spread of Dermo have not clarified the exact conditions that led to its initial appearance in the Northeast. There is speculation that Dermo may have arrived as a result of the introduction of southern oysters. Such an event is believed to be the cause of the initial Dermo infection in Delaware Bay in the 1950s, when Chesapeake oysters were imported for planting. Currently, Dermo may be enzootic; there is evidence that it exists at low levels within populations throughout the Northeast but does not always cause massive outbreaks. Perhaps high summer temperatures promote its growth and infectivity. The Northeast epizootics also may be the result of the emergence of novel, cold-tolerant strains of *P. marinus* (Ewart and Ford 1993).

Similar to MSX, the disease is prevalent in warm, saltier water (see table 13.1). Infection causes the oyster meat to become thin and watery. In contrast to the earlier summer arrival of MSX in the northeast, Dermo is more prevalent in September and October. In 2006, Dermo was associated with late-fall mortality of farmed oysters in Wellfleet, Massachusetts. Also in contrast with MSX, Dermo appears to be transmitted directly from oyster to oyster by means of motile spores. Dead and dying oysters are also sources of infective *P. marinus,* which can be spread by scavengers of dead oysters. In the southern part of the Unites States, a snail, *Boonea,* has been identified as a vector, spreading Dermo to oysters as it feeds. Direct transmission can be demonstrated under laboratory conditions. Infection of oysters occurs after the parasite is ingested and passes through the stomach or intestine.

Scientists have been working on the development of Dermo-resistant oyster lines and several have been developed, including the DEBY oyster line, which was developed by breeding survivors of Delaware Bay Dermo epizootics and the NEH line developed at Rutgers. Evidence suggests that native oyster survivors from other areas with heavy Dermo outbreaks display significant Dermo resistance. The most important measures to control Dermo are to prevent the introduction of infected organisms into areas where oysters are not infected and to prevent movement of uninfected oysters into areas where Dermo may be present. Cold winter temperatures usually are associated with a decrease in the spread of Dermo, but the disease may make its appearance again during the following spring. Summer survivors of Dermo may not be able to tolerate the parasite after emerging from winter dormancy. Many an aquaculturist has breathed a sigh of relief when there were no losses to Dermo in the fall, only to find significant mortality during the second year of growth. In the Northeast, a mild winter is sometimes a contributing factor to the incidence of the disease.

Even though MSX and Dermo are different diseases, caused by different organisms, some general measures can be used to decrease the prevalence and prevent the spread of these diseases. Diagnosis of the disease in a timely manner can be a great aid in the management of shellfish beds. Shellfish farmers should consider prompt removal of all visibly diseased and dying oysters from infected areas and harvest other oysters (which probably are infected but asymptomatic) for market before they succumb to the disease. In addition, the infected area should be left fallow for at least a year. It may be possible to practice a type of crop rotation in which aquaculturists alternate oyster crops with hard clams. In addition, special vigilance should be maintained during drought periods, when water usually becomes more saline. Communities where shellfish harvest and/or aquaculture are important commercial ventures should manage their resources to ensure that freshwater flow to shellfish areas is not restricted.

JOD

JOD, juvenile oyster disease, typically affects younger oysters, less than 25 mm (about 1 inch) in height, and is promoted with high salinity. The diseased oysters are most affected in mid- to late summer when water temperature is at its seasonal high. Affected oysters display uneven shell growth and the appearance of "brown rings" caused by deposits of conchiolin on the inner shell. Outbreaks of JOD have occurred in hatchery stock, grown at high density. *Roseovarius crassostreae* has been identified as the bacterium responsible for this infectious disease. The problem has not been seen in wild oysters and the reasons for its appearance in the hatchery are unknown. The disease is inhibited if the young oysters are grown under low-salinity conditions. JOD can be controlled by decreasing the density of oysters and/or increasing water flow through nursery trays or upwellers.

SSO

SSO or Seaside organism, *Haplosporidium costale,* causes problems with condition and mortality similar to Dermo and MSX. Microscopically, the organism resembles *Haplosporidium nelsoni.* SSO received its name because it was first discovered on Virginia's ocean coast. It has been found in the mid-Atlantic and in the Northeast, but so far has not had as large an impact as MSX and Dermo. Similar to the other oyster diseases, high salinity is a contributing factor to its prevalence.

Bonamiasis

It is also worth noting another disease of oysters, bonomiasis, caused by *Bonamia ostreae.* This disease is not seen in eastern oysters, but can be a problem for Northeast aquaculturists who are growing European flat oysters, *Ostrea edulis* (Gosling 2003). Similar to the organism that causes MSX, bonamiasis is spread by a haplosporidian protozoan in the Phylum Acetospora. The disease has moved across the Atlantic along with its oyster host. It made its first U.S. appearance in California and was then moved back to Europe via seed oysters that were sent to Brittany, France. In Europe, it caused complete devastation of cultured as well as natural stocks of *O. edulis.* Today, bonamiasis is found predominantly on the West Coast of the United States, but it has been detected at low prevalence in Maine as a result of oyster introductions prior to 1970.

The disease is transmitted from oyster to oyster. After entry of the infectious organism, host blood cells, called hemocytes, engulf it but cannot destroy it. *B. ostreae* multiplies within the hemocytes and spreads throughout its host. Many organisms are affected and mortality ensues. Not much can be done by way of prevention. If bonamiasis has made its appearance, the farmed area should be left fallow for several years. In France and Ireland, strains have been developed that are relatively resistant to the disease, so that oysters can be grown to market size before they display symptoms (Gosling 2003).

Diagnosis of Oyster Diseases

The appearance of diseased oysters does not always indicate the identity of the parasitic culprit. Under most circumstances, when a diagnosis is important, dead and dying animals are sent to a veterinary pathologist for detection of microbial parasites and laboratory analysis of tissues from the bivalves. Histological examination is the traditional manner for diagnosis of MSX. Dermo can be diagnosed using the Ray/Mackin tissue assay. Tissue from the rectum and gill or mantle is excised from the oyster and placed into Ray's fluid thioglycollate medium (RFTM). The samples are incubated for five to seven days at 25 to 30°C (77–86°F) under conditions without oxygen. Samples are then stained with Lugol's iodine solution and examined under the microscope for the presence of structures known as hypnospores, characteristic of Dermo. The RFTM test is time-consuming. Neither the RTFM test nor histological examination can detect MSX or Dermo organisms in the environment.

It is increasingly important for the success of shellfish aquaculture that these diseases be diagnosed in the early stages of infection. For example, hatchery seed can be screened for the presence of the parasites so that infected seed is not purchased by growers nor used in locations where the disease is not found. It also may be important to diagnose the presence or abundance of the parasites in the environment. This information can be used to guide management strategies in shellfish-growing areas.

The development of specific, sensitive tests to detect MSX and Dermo in the very early stages of infection or in the environment has proved challenging. Scientists have been pursuing a new strategy for detecting these disease-causing microbes. Rather than looking for the microbe itself either by direct microscopic observation or by culturing it in the laboratory, a shellfish pathologist might look for traces of the microbe's DNA, or its molecular footprint. In order to do this, scientists must use appropriate probes, which are DNA segments that mirror or complement specific DNA segments from the microbe. If the probe is specific, it will detect only the microbe of interest and not its close relatives. Theoretically, the presence of a single microbe can be detected with the probe by polymerase chain reaction (PCR) because the technique amplifies the DNA of the microbe so that millions of copies are made. The microbe's footprint is then detectable with laboratory instruments. Stokes, Siddall, and Burreson (1995) used a sensitive PCR assay to detect *H. nelsoni* in hemolymph or tissues from oysters that tested negative with traditional histological methods. PCR can be modified to quantitate the amount of microbe DNA in a sample so that relative amount of parasite can be known, rather than simply presence or absence. Day, Franklin, and Brown (2000) modified the PCR reaction to allow the quantitation of *H. nelsoni* in a technique called quantitative-competitive PCR (QC-PCR).

Once a probe has been developed and found to show specificity in the laboratory, it can be field-tested to demonstrate its validity as a diagnostic tool. Using a Dermo-specific probe, Audemard et al. (2006) demonstrated the presence of *Perkinsus marinus* in wild oysters and in the water column in Chesapeake Bay. The presence of the parasite in the water and in oysters was shown to increase with water temperature. Salinity also positively correlated with the dynamics of infection. The study demonstrated the positive relationship between oyster mortality and abundance of *P. marinus* in the water column, supporting the idea that dead oysters release the parasite. Their study also suggested that *P. marinus* is shed in the feces of infected oysters.

A promising variation of the PCR method uses probes for more than one microbe in the same test. This technique, known as multiplex PCR, has been developed for detection of MSX, Dermo, and SSO in a single sample (Penna, Kahn, and French 2001; Russell et al. 2004). As the molecular tools are refined, and as the molecular tests become standardized and more widely applied, it will be possible to learn more about the presence and transmission of the parasites in nature, refine models to predict the spread of the diseases, and develop management techniques to mitigate losses. We also will be able to learn more about the life cycle of these fascinating but deadly parasites and how host bivalves are responding to them and developing strategies to survive.

Selective Breeding for Disease Resistance

Scientists have observed that oysters that survive an outbreak of MSX, Dermo disease, or bonamiasis tend to be those that have some resistance to the offending parasite. These disease-resistant bivalves become infected with the parasite but often reach maturity, and are often harvestable, before the parasite induces serious disease symptoms. For the aquaculturist, the disease-resistant individuals represent the most desirable segment of the variation that occurs in natural populations. The presence of the disease itself in the population is selecting for better survival in the most resistant individuals. The availability of laboratory-bred disease-resistant lines in addition to naturally resistant native bivalves provides sources of broodstock (Calvo, Calvo, and Burreson 2003; Guo et al. 2003). These stocks can be used to produce bivalves in hatcheries and have the potential to aid aquaculturists who farm in areas where the diseases have been a problem.

In most cases, the reason for the resistance is not known. Scientists are comparing the immune function of disease-resistant organisms to that of susceptible animals. An intriguing idea that may explain resistance to Dermo in oysters can be traced to the activity of hemocytes, an immune system cell found in hemolymph. Hemocytes are capable of ingesting *P. marinus* in a process known as phagocytosis. One method the immune system employs to rid the organism of the pathogen is to promote the destruction of the actual host hemocytes that contain the pathogen, a biological process known as apoptosis, also known as programmed cell death. It's a type of suicide mission that kills the hemocyte but also kills the pathogen. *P. marinus* has the ability to suppress apoptosis of hemocytes in susceptible oysters. Goedken et al. (2005) have found that Dermo resistance correlates with an increased ability to overcome the suppression of apoptosis by *P. marinus.* Thus, apoptosis may have an important function in combating Dermo.

Quahog Diseases

Mats of dying quahogs, rising to the surface with their shells agape, spelled trouble to shellfish farmers in Provincetown, Massachusetts, during the summer of 1995. When the die-off problem was first noticed in the early 1990s, it originally was attributed to crabs, due to the chipped edges along the margin of the clam shells. As the die-off became more extensive, it took on the features of a disease that was spreading slowly. Further investigations by scientists at Woods Hole pointed to a strange single-celled, fungus-like organism belonging to a group known as a thraustocytrids. Thraustocytrids are protists, organisms that belong to the Kingdom Protista. This large kingdom also includes other single-celled organisms such as amoebae and paramecia and the protozoans that cause the oyster diseases, MSX and Dermo. The organism that was implicated in the Provincetown clam mortality belongs to a poorly characterized group (Phylum Labyrinthomycota or Labyrinthomorpha; Family Thraustochytridiae) with the common name slime net protists, found in marine and estuarine environments. The name comes from the observation of microscopic filaments (ectoplasmic nets) that are extruded from the organisms and help them to attach to substrates. The Provincetown pathogen may be more primitive than other members of the Thrausochytrid family.

The Provincetown organism was shown to be similar to one that caused clam mortality in Canada in the 1950s and 1960s, and at a hatchery in Prince Edward Island, Canada, in 1989, when it was given the name QPX, Quahog Parasite Unknown. In one study, the filaments of QPX were characterized by histological staining as a mucofilamentous net, indicating that the chemical compostition of the nets may consist predominantly of mucopolysaccharide (Smolowitz, Leavitt, and Perkins 1988).

At about the same time that clams were dying in Provincetown, across Cape Cod Bay the same disease hit in the town of Duxbury, Massachusetts. Sporadic clam-mortality events in Barnegat Bay, New Jersey, in 1976, and in Chatham, Massachusetts, in 1992, most likely were caused by QPX or a QPX-like organism (Leavitt and Crago 1996; Ford 2001). The Provincetown quahog farming industry has not yet recovered from its destruction by QPX in 1995.

QPX or Chytrid-like disease can cause significant mortality of quahogs. Once the disease has made its appearance, the infectious organism may remain in the water column or sediments for long periods of time. Current thinking proposes that QPX may be a "facultative" organism. This means that it is not an obligate clam parasite; it does not depend on the clam for its survival or reproduction. Under the appropriate circumstances, the organism can infect and colonize clam tissues and cause disease. It is not known whether QPX itself can infect organisms other than quahogs. Other Thraustochytrids, related to QPX, are known to cause diseases in squid, octopus, and sea grass.

Although QPX may have been present in Wellfleet waters for many years, the first massive QPX infection in Wellfleet was reported during the winter of 2004. It took everyone by surprise. Dead, gaping clams were discovered in a small area of Wellfleet Harbor known as Egg Island and appeared to be restricted to areas licensed to three shellfish aquaculturists. Shellfish pathologist Dr. Roxanna Smolowitz and Barnstable County Cooperative Extension aquaculture specialist Bill Walton were brought in for quahog triage. Gross examination revealed the hallmarks of QPX: chipped shells, small nodules, and meat retracted from the shells. Later, microscopic examination confirmed the diagnosis.

Microscopic analysis of QPX in diseased clam tissues reveals three forms of the organism. Small, round bodies, representing the vegetative (nonreproducing) stage of the organism, are approximately 2 to 20 μm in diameter, often with a clear area or halo surrounding the organism (Smolowitz, Leavitt, and Perkins 1998). Another form involved in the reproduction of QPX is known as a sporangium (plural sporangia). It is 10 to 48 μm in diameter and contains 20 to 40 spores, each measuring 2 to 5 μm. When sporangia rupture, the spores are released, with each spore potentially giving rise to another vegetative cell. A motile spore is most likely the form of QPX that passes from clam to clam and is implicated in the spread of the disease. Thus QPX is transmitted directly without the need for an intermediate host. Under controlled laboratory conditions, Smolowitz et al. (2001) showed that healthy clams could be infected with QPX after three months of exposure to QPX-infected clams. In support of the direct-transmission hypothesis, preliminary work from the Marine Disease Pathology and Research Consortium in New York suggests a direct transmission of QPX to healthy clams from QPX grown in laboratory cultures.

QPX is found in many marine environments and prefers high salinity. On shellfish beds, clams infected with QPX may be able to spread the organism to nearby clams. However, QPX transmission is not a straightforward chain of events. A number of factors must come into play, including growth conditions such as density of clams. Other factors such as food limitation can contribute to stress and put clams at risk of infection. Environmental factors such as water temperature, salinity, and numbers of infectious organisms in the water and sediments are also important. The mucoid net produced by QPX may interfere with clam immune defenses. Differential host susceptibility has been demonstrated experimentally. During a three-year study, clams were taken from stocks in five locations in Massachusetts, New Jersey, Virginia, South Carolina, and Florida. The clams were grown in New Jersey and Virginia in tests for QPX susceptibility. Clams from the five locations also were transplanted to Massachusetts but suffered such high initial mortality that they could not be included in the study. Results of the experiment revealed that southern clams—from Florida and South Carolina—proved much more susceptible to QPX than those from northern stocks—from Massachusetts, New Jersey, and Virginia (Calvo and Burreson 2002). Another experiment involved the direct inoculation of three different strains of QPX into the pericardial cavities (heart region) of quahogs from stocks obtained at four different geographical locations in Massachusetts, New York, Virginia, and Florida. Tolerance to QPX displayed a latitudinal pattern, with the northernmost clam stocks displaying the most tolerance while the Florida strains had the highest infection rate (Dahl, Perrigault, and Allam 2006). These results suggest a genetic component to QPX susceptibility, a fact that should be of interest to shellfish farmers who buy quahogs from nursery stocks. The geographic origin of the stock may be a key factor in the success of a shellfish farm if QPX is in the neighborhood. It is still a mystery why one bed of cultured clams can have severe QPX mortality while a nearby bed does not show signs of disease. The mystery may be partially explained if seed clams came from different geographic sources.

In addition to the susceptibility of quahogs from different stocks, the virulence of the QPX organism itself plays an important role in the disease process. Different strains of QPX may vary in their structure and/or in their ability to cause disease. For example, the strain that caused QPX outbreaks in Canada in the 1990s was not identical histologically to the one that caused the Wellfleet outbreak. Not enough is known about QPX to posit the existence of forms that are genetically different, however, isolates from different areas appear to display different growth characteristics when grown in laboratory cultures.

No matter the "strain" or the exact mode of transmission, the interim diagnosis of QPX in Wellfleet raised alarms. Massachusetts Extension Agent Dr. Bill Walton described the outbreak as a "Black Plague of clams." Under the direction of Shellfish Constable Andy Koch, the Wellfleet aquaculturists mobilized. Within a short period of time, volunteers assisted the growers in the infected areas as they dug up their clam plots, shoveled the dead and dying clams into a dumpster and disposed of them at the town landfill. Because of this quick response, masssive spread of QPX to other areas of the harbor was averted. During the clean-up phase of the operation, the shellfishermen also were urged to use plenty of freshwater to rinse all gear and clothing that came in contact with the sick clams and the area that was cleaned out, because QPX does not survive in freshwater. Freshwater rinsing can prevent the disease from being spread indirectly and inadvertently to other locations. In Rhode Island, wild populations of quahogs have tested negative for QPX for five years (2000 to 2005), but QPX outbreaks have occurred in farmed quahogs in certain locations.

QPX put the brakes on a shellfish transplant program in New York. Since 1987, the New York State Department of Environmental Conservation (DEC) has moved clams from Raritan Bay, off Staten Island, to growing areas in other parts of the state. The Raritan Bay clams are not fit to eat; the waters around New York City are "uncertified" for harvesting shellfish. The clams are moved to Peconic Bay, Long Island, and other relatively clean New York State locations during the growing season when the clams can filter the cleaner water and pump out any Raritan Bay pollutants. After three weeks in clean water, the clams can be approved for harvest. This relay is part of an important $5 million commercial fishery that supplies about 45 percent of New York's hard-clam production (New York State Department of Environmental Conservation 2005). But QPX was found in Raritan Bay clams in 2002, the first time it was ever discovered in New York. The transplant program was shut down to prevent transmission of QPX to wild Peconic Bay clams. An alarmed U.S. senator from New York, Charles Schummer, wrote to the National Marine Fisheries Service requesting aid in seeking another source of unmarketable clams that could benefit from a three-week bath in Peconic Bay. After close to three years of study and monitoring, a decrease in prevalence of QPX led to a reopening of about 25 percent of the Raritan Bay flats to the transplant program. Vigilance and continued monitoring will be important aspects of the management of this program to prevent a QPX epizootic in the hard clams of Peconic Bay.

Without a "cure" for QPX, the best method to prevent the disease in farmed shellfish involves the close scrutiny of seed stock. Monitoring the geographic origin of the seed and testing seed stock for presence of QPX before planting it on shellfish beds are of primary importance. Another preventative action involves planting clams at lower densities to avoid stress and competition for food. Once the disease has made its appearance in an area, the Wellfleet response model could be considered. Rapid removal of all clams (whether or not they show signs of disease) from an infected area may halt the spread of QPX and prevent heavy contamination of the area and of nearby quahog farms.

Recent Research in QPX Disease

On a beautiful fall day, I traveled to Woods Hole, Massachusetts, to visit Dr. Roxanna Smolowitz in her laboratory in the Marine Research Center at the Marine Biological Laboratories. I heard her speak about shellfish disease at a Cape Cod Natural History Conference, so I knew she was performing important research in the field. Dr. Smolowitz, or Rox as she is known to colleagues, is an expert on shellfish ailments. She became interested in this field from her early years as a veterinarian when she studied marine and exotic animals. She is an aquatic veterinarian with a specialty as a veterinary pathologist. She uses techniques such as gross observation, tissue analysis using microscopic techniques, and immunological methods, and is collaborating with other researchers on the employment of the most modern molecular methods such as PCR. QPX molecular diagnostic methods have been developed at Virginia Institute of Marine Science in Eugene Burreson's laboratory (Stokes et al. 2002). With molecular tools, Smolowitz can diagnose a shellfish disease and determine its severity. But her goal is not simply to be able to diagnose a shellfish disease. She has embarked on many studies to learn the causes and modes of transmission of these diseases as well as possible ways to prevent them.

One of the reasons I met with Smolowitz was to learn about the current status of the newly developed molecular tests to diagnose shellfish diseases. I wanted to find out if the tests could be used routinely to monitor shellfish areas for the presence of the disease-causing organisms. Not only did I get an update on the tests, I learned a great deal about the complicated nature of these diseases and that there was much to be learned before the tests could be performed in an accurate way to predict the occurrence or spread of the pathogens.

One of the complications with environmental testing is the fact that disease-causing organisms can be found in sediments whether the resident shellfish have the disease or not (Lyons et al. 2006). The presence of the potential pathogen in the sediments or the water column may not be predictive of a disease occurrence. It's also not clear whether the amount of the agent present is correlated with the susceptibility to the disease.

In shellfish beds, the time between initial infection and full-blown disease may vary greatly in individual animals. For example, QPX may manifest itself just as clams become market size and are ready to be harvested. Smolowitz cited the fact that larger animals filter more water and thus have the potential to take in more pathogens. They also have a larger mass of tissue, so it may take longer for the condition of the tissue to deteriorate. Another variable is the different susceptibility to diseases in different genetic strains of the same species.

Smolowitz has "poured" QPX into tanks containing quahogs. Simple exposure of clams to cultures of QPX in the laboratory does not lead readily to QPX disease. Direct injection of the organism is often necessary to initiate disease symptoms. This suggests that clams can handle or eliminate the QPX organism in most cases. Smolowitz and her colleagues at the University of Connecticut and Stony Brook University in New York have been studying whether an important factor in infection may be the manner in which the clam is exposed to QPX. Working under this hypothesis, they have designed a simple collection device to sample the water column for marine aggregates. These tiny, sometimes microscopic particles are also known as flocs, floating detritus, and marine snow. Using molecular techniques, Smolowitz and colleagues have confirmed the presence of QPX in marine aggregates in Massachusetts waters. They also were able to confirm that these aggregates, which could contain QPX, entered the pallial cavity of clams (Lyons et al. 2005). Smolowitz hypothesizes that if the marine aggregates are not digested, they may accumulate in pseudofeces and thus be the source of infection. The group has developed a quantitative PCR-based molecular test that can be used to improve diagnosis and help to characterize the levels of the organism in the environment (Lyons et al. 2006).

Smolowitz can diagnose QPX in clams without the need for sophisticated molecular probes. Since QPX initially accumulates in the siphon of *Mercenaria,* a microscopic view through a slice at the base of the inhalant siphon tissue can quickly reveal the presence of the organism. Currently, the diagnosis of diseases of oysters with molecular tools may be more efficient than the diagnosis of QPX organisms in clams using similar molecular assays.

Cancer

Bivalves are not immune from cancer. Various types of neoplastic (cancerous) diseases have been described in hard and soft clams, mussels, and oysters. Some types of cancer involve the hemic system (blood and lymph) and are collectively called disseminated sarcomas because they don't form a single discrete tumor; other cancers are specific to gonadal tissue (Ford 2001). The Registry of Tumors in Lower Animals (RTLA) at the Smithsonian Institution in Washington, D.C., has samples from over fifty documented cases of bivalve cancer, mostly from the United States. From the Northeast, the list includes tumors from ocean quahogs (*Arctica islandica*) from Rhode Island Sound, eastern oysters (*Crassostrea virginica*) from Searsport, Maine, and New Haven Harbor, Connecticut, and cancers of softshell clams (*Mya arenaria*) from several sites in Maine, Boston Harbor, and other locations in Massachusetts and Rhode Island (Peters 1988).

The disseminated cancers were first noticed in tissue sections after histological preparation and microscopy. Histopathology is still used to diagnose the condition, although immunological techniques can be used to distinguish diseased cells from normal ones. Most bivalve cancers are still poorly or incompletely understood and the exact cell type of origin is not always known with certainty. A variety of cellular proliferative diseases may appear similar or identical. Although many are due to proliferation of hemocytes, not all can be traced to this cell type. Hence, a variety of terminology has appeared for the disseminated neoplasms. They have been called "hematopoietic neoplasms, diffuse neoplasm of hyaline hemoctye origin, diffuse sarcoma of possible amoebocyte stem cell origin, hemocytoblastic sarcoma, hemocytosarcoma, disseminated hemic sarcoma," also, "undifferentiated sarcomas, poorly differentiated sarcoma of undetermined origin, diffuse mesenchymal sarcoma" (Peters 1988, 75).

A disseminated neoplastic disease of softshell clams has been studied extensively in the Northeast. The term "clam leukemia" was coined by Carol Reinisch and her colleagues at Tufts University and the Marine Biological Laboratory at Woods Hole. The cancer cells of the clam express the same tumor genes that are seen in human leukemias (Gosling 2003). The stages of clam leukemia have been characterized. Beginning with the initial proliferation of abnormal hemocytes, first neoplastic hemocytes spread throughout the soft tissues. The neoplatic hemocytes fill the vascular spaces and stop the flow of hemolymph. Eventually, the clam will die, although some reports of remission appear in the scientific literature.

Despite the fact that the exact cause of clam leukemia is not known, one of the most interesting aspects of the disease is that it is transmissible. A virus, possibly a retrovirus, has been implicated. Pollution also has been associated with occurrences of leukemia in clams. Contaminants such as polychlorinated biphenyls (PCBs) and polycylic aromatic hydrocarbons (PAHs) have been found in the water and in clam tissues where outbreaks of this type of cancer have occurred. However, conflicting laboratory and field findings have been reported. Cancers have been found in clams in pristine sites and exposure of clams to pollutants in the laboratory does not always lead to the development of neoplasm. It is likely that a combination of conditions leads to increased susceptibility to the disease.

Because of the transmissible nature of softshell clam leukemia, it is recommended that: (1) affected clams should not be introduced into area where the disease does not exist; (2) softshell clams should be grown at low density; (3) clams should be harvested before the peak mortality period (usually fall and winter); and (4) clams should be harvested as young as possible at legal size because susceptibility to cancer increases with age.

Quahogs also may be susceptible to cancer. Veterinary pathologists have found leukemia cells in quahogs from aquaculture operations; however, no one has reported mortality attributed to cancer in quahogs. Curiously, hard clams may produce anti-cancer compounds such as "mercenene" and an unnamed peptide isolated from Asian hard clams, *Meretrix meretrix,* which has been shown to inhibit the growth of human gastric gland carcinoma cells (Schmeer, Horton, and Tanimura 1966; Leng, Liu, and Chen 2005).

These complex diseases that can afflict shellfish warrant further study. They have the potential to cause serious economic problems for growers and make potential newcomers to the industry think twice about making a steep investment in time and capital to start a shellfish farm. The prevalence of these conditions is certainly an incentive for some growers to diversify their crops or have a backup plan to make a living when their shellfish crops are stricken.

Chapter 14

Foul Play

Predators, pathogens, and parasites affect the health of shellfish by specific mechanisms and interactions. However, nonspecific interactions with other organisms also can affect the health and survival of bivalves. For example, algae that grow in large mats can smother bivalves by depleting oxygen from the water and starve them by preventing the growth of the phytoplankton that nourish them. Organisms that grow on or foul shellfish gear can increase the work of the farmer and may even cause loss of the crop.

Many organisms reside in coastal ecosystems and share habitat with cultured shellfish. Some of the organisms are beneficial or even a necessity for the shellfish farm, such as the phytoplankton that bivalves filter and feed upon, but others can impede the growth of bivalves and have a negative impact on the economic return for the shellfish farmer. These organisms do not cause disease and hence are not classified as pathogens or parasites. Some simply compete for food; others compete for surface to grow on or space to spread. Some impede the flow of water and/or oxygen that directly affects the availability of phytoplankton, the capacity to remove waste products, and the ability to breathe. A few organisms can cause trauma to a bivalve's shell. Some pesky organisms have multiple minor effects that, when taken together, are detrimental to shellfish farms.

These pests, summarized in table 14.1, not only grow on, in, and around shellfish, they also grow on, in, and around shellfish gear and cause a condition known as "fouling." The gear sits in the water for extended periods. Like rocks along the shoreline or manmade structures such as piers, docks, and pilings, or even the bottoms of boats, living things will eventually attach and accumulate (fig. 14.1). These collections of flora and fauna make perfect specimens for identification of marine creatures by biology students. More and more, the identity of the attached organisms can be traced to nonnative, alien, or invasive species. Such alien invasions have been identified as a major ecological problem in maintaining biological diversity

Table 14.1
Common Shellfish Competitors, Pests, and Fouling Organisms

Organism	Effect
Boring sponges (*Cliona* species)	Excavate within the shell where it lives, create cavities that eventually may weaken the shell
Mud worms, also known as blister worms (*Polydora* species)	Bore into oyster shells, produce blisters that become filled with mud
Jingle shell (*Anomia simplex*)	Grows over oyster spat
Slipper shells (*Crepidula* species)	Set at same time as oyster spat, grow faster than oysters and compete for space and food
Blue mussel (*Mytilus edulis*) (not a pest for mussel growers!)	Competes with oysters for food and space, blocks water flow into oyster cages and trays, weighs down gear
Barnacles (*Balanus* and *Semibalanus* species)	Compete for food and setting space
Tunicates (sea squirts) *Styela* (club or stalked tunicates), *Botryllus* (star tunicates)	Attach to gear and hard substrates, weigh down gear, and decrease water flow
Sea grapes (*Molgula* species)	Attach to gear, weigh down gear, and decrease water flow
White crust (*Didemnum* species)	Carpets sea floor, especially if the substrate is gravel or pebble; can smother shellfish
Hydroids (*Tubularia* and *Campanularia* species)	Attach to gear and hard substrates, impede water flow
Tube worms (*Hydroides dianthus* and *spirorbis* species)	Attach to hard surfaces and plants
Bryozoans	Common space competitors. Cover oyster shells with their colonies and prevent spat settlement
Algae	
Rockweed (*Fucus* species)	Attaches to bivalves and gear
Sea lettuce (*Ulva* species)	Attaches to bivalves and gear; can form smothering mats
Hollow green weeds (*Enteromorpha* species)	Become entangled in gear
Codium fragile spp. tomentosoides Aka dead man's fingers, fleece alga, Sputnik weed, broccoli weed, oyster thief	Attaches to bivalves and gear, can collect silt and cover oyster beds

FIG. 14.1 Fouling of shellfish gear. *Photo: Captain Andrew Cummings*

of natural marine ecosystems; the aliens sometimes grow faster and out-compete the native organisms.

The alga *Codium* provides a good example of the impact of an invasive species on shellfish beds and aquaculture operations (plate 17). *Codium* is native to the Asian Pacific region. It has been dispersed as a result of human activities such as boating and transport of shellfish. It was first discovered in the Northeast in 1957 in Long Island Sound and also near Montauk Point. In 1961–1962, it was found on Cape Cod near Chatham, where it may have been carried along during an oyster restocking effort. It is believed to have drifted to Rhode Island shortly after and was found on Moonstone Beach in May of 1962 (Wood 1962). *Codium* can cause serious economic losses to shellfish operations, hence the name "oyster thief." Globally, it is one of the most invasive algal species. It hinders feeding of bivalves, movement of scallops, and can displace eelgrass.

Fouling organisms can coat the bottom like a carpet and prevent the settlement of oyster spat or smother quahogs growing a few inches beneath. In addition to their effects on naturally occurring bivalves, these organisms are also a nuisance for the shellfish farmer. They weigh down the gear and make it difficult to haul or move equipment. More importantly, heavy fouling of gear will impede water flow to the growing bivalves, cutting off their supply of food and oxygen. The fouling organisms that compete for food cause decreased growth rates in cultured bivalves. Not all fouling organisms are invasives. In areas where oysters spawn, spat may attach to shellfish bags or to the cultured oysters themselves (a condition known as overset) and, as they grow, prevent the growth of the cultured oysters within the bags. Some shellfish operations focus on the commercial importance of the blue mussel, *Mytilus edulis,* but most oyster growers consider mussels to be a serious enemy. Mussel cultivators also have problems with fouling. Mussels often are cultured on long-lines, suspended in the water. The same parameters that promote the settlement and growth of mussels favor the attachment of all sorts of marine life.

The green crab, *Carcinus maenas,* is an invasive species. This small crab may be a nuisance when it preys upon tiny bivalves. However, when oysters grow larger than the crab, there may be a role reversal. The potential benefits of green crab were demonstrated to me at the East Dennis Shellfish Farm, operated by John and Stephanie Lowell on a pristine stretch of Cape Cod tidelands near Crowe's Pasture Conservation Area. The sign at the entrance indicates the concerted effort of the various agencies that established and maintain the area: the Town of Dennis/Land Bank project, Dennis Conservation Trust, U.S. Fish and Wildlife Service, Federal Land and Water Conservation Fund, Massachusetts Department of Conservation and Recreation, and Massachusetts Self-Help Program. This is state-owned land and without the foresight and planning by these agencies, this stretch would be studded with trophy homes. The East Dennis oyster farmers don't have to worry about complaints from upland property owners.

John Lowell was kind enough to show me his farm and walk me through some of the ways he labels and tags his oysters so they can be identified as they move from the flats to the plates of consumers. As John opened a tray and offered me an oyster to sample, he showed me his little helpers: small green crabs that he finds on the flats and puts in each oyster tray. I was somewhat shocked. I had been told that crabs are the oyster's enemy. But under the controlled conditions of the oyster farm, the crabs can be put in the trays after the oysters are too large for them to prey upon. John and other growers notice that the crabs within the trays are getting bigger, so they must be eating something. It is also apparent that the amount of fouling on John's trays and bags is minimal. They look picture perfect, like they had just been placed on the farm. The working hypothesis is that the crabs are eating tiny barnacles, mussels, and other organisms that otherwise might settle and foul the trays.

Some aquaculture sites in intertidal locations are deprived of water for several hours each day. Cultured oysters can tolerate being high and dry under these conditions but many fouling organisms are sensitive to desiccation. A licensed shellfish area that experiences dry periods and huge tidal washes may remain relatively pristine throughout the growing season. Of course, there is a trade-off. If oysters are deprived of water flow for several hours a day, they are also deprived of their phytoplankton meal and may not grow as fast as oysters in other areas.

In deep-water aquaculture areas, bivalve farmers may spend a good part of the summer maintaining gear and removing fouling organisms. Shellfish trays, cages, and bags may have to be scrubbed and rinsed throughout the growing season. If the aquaculturist has sufficient extra equipment, the bivalves can be transferred to clean trays, cages, or bags and the fouled equipment can be brought to shore and air-dried or power-washed. Gear usually is hauled in by boat, washed, and returned to the licensed area. Sometimes, the power-washing is done at the aquaculture site. This practice may be convenient for the shellfish farmer but may pose ecological problems, especially when it enhances the spread of invasive organisms. Environmentalists discourage power-washing of gear at the aquaculture site and some programs are helping shellfish farmers to purchase extra gear so they can transfer their oysters at the site, deploy fresh gear, and bring the fouled gear to shore for washing.

Another strategy to mitigate against fouling is to employ gear that is treated with anti-fouling compounds. Use of gear treated with anti-fouling materials can save the shellfish farmer the time and labor required to maintain equipment and can reduce the cost of aquaculture operations. Unfortunately, many anti-fouling compounds, such as those used to paint boat hulls, have been shown to have adverse environmental effects. Although they can be effective, they are relatively nonspecific and can be toxic to shellfish, especially juveniles. Using a novel approach, some companies are studying marine creatures that may be producing substances that prevent other marine organisms from adhering to them. If such substances can be identified and either extracted or synthesized in the laboratory, they might be incorporated into paints, plastics, ropes, and other aquaculture equipment.

A number of the organisms that have been identified as predators, pests, or fouling organisms in shellfish aquaculture operations are normally quite harmless in the wild and some have important ecosystem functions. For example, algae serve as sources of food or habitat for other organisms. Algae become problematic when conditions are unbalanced. When excessive nutrients are loaded into water bodies, algae can overgrow, which can result in decreased light penetration, thus having an effect on the growth of plants such as eel grass (*Zostera marina*), an important nursery habitat for juvenile scallops (McGlathery 2001). After large mats of algae grow, they eventually will die and sink to the bottom. This creates anoxic conditions, an additional detriment to bivalve health, because microbes use up the available oxygen as they decompose the algae.

Before extensive harvesting, disease, and pollution took their toll, oyster beds covered the bottoms of Northeast tidelands and scallops were abundant in deeper waters close to shore. The former abundance of quahogs was documented by early settlers in New York and New England. The bivalves did very well despite the presence of many creatures currently described as predators, pests, and fouling organisms. Today's serious environmental microbial pathogens, MSX, Dermo, SSO, and QPX, were not a factor; they were brought into play by human activities as bivalves were moved around and planted in foreign tidelands.

Now, we are full tilt into management of our shellfish resources. Without management, no shellfish will appear on the menu or in the seafood market or seafood section of our supermarket. As we manage our shellfish on farms and on remaining natural beds, we also must learn to understand and control the organisms that we have come to regard as enemies: predators, parasites, pathogens, and pests. But we should remember that some of these creatures have a place in the ecosystem and all-out warfare might be inappropriate. It would be ideal if we could recreate the balance that existed prior to European colonization, between predator and prey, pest and host, in order to maintain the health of our shellfish and also the diversity of organisms that nourished the original human inhabitants of the Northeast.

Chapter 15

Celebration of
the Hard Harvest

During the 2006 Oysterfest in Wellfleet, Barbara Austin, the only woman contestant, outshucked her formidable opponent by a mere second, to take home the $1,000 prize, a portion of which she donated to a local charity. The Wellfleet contest qualifies the winner to enter the National Shucking Contest in Maryland and is one of the nearly twenty such shuck-offs globally. To claim the title, Austin shucked two dozen Wellfleet oysters in the adjusted time of 3 minutes, 52 seconds. The previous year, winner William "Chopper" Young shucked his 24 oysters in 1 minute, 28 seconds, and went on to win fourth place in the nationals. The Wellfleet competition inaugurated Chopper into the world of shucking competitions, and he has been bringing home national and international prizes since his Oysterfest debut. Speed is not the only factor in this event; Austin's time was adjusted due to penalties set by the judges for presentation of the oysters. Factors such as broken shells, shell fragments in the meats, cut meats, and blood on the oysters (shucking can be dangerous, especially if contestants do not wear gloves) can add seconds to total shucking time. During the 2007 Oysterfest, Austin defended her title while improving her time to 2 minutes, 51 seconds. Contestants take their work seriously; many wield custom knives that perfectly fit the contours of their hand and have the proper strength, sharpness, and shape of the blade.

Wellfleet's Oysterfest is a celebration of the bivalve that was inaugurated in 2001 by a group of Wellfleetians who subsequently formed an organization known as SPAT: Shellfish Promotion and Tasting. SPAT sponsors events and seminars and provides scholarships for local youth in efforts to promote interest in the town's shellfishing traditions and appreciation of its rich history as a shellfishing community. SPAT's planned celebration originally was intended to be a small, local venue, attended by a few hundred people, but by 2006, Wellfleet Oysterfest had mushroomed to an oysterama enjoyed by an estimated twenty thousand visitors, creating the human counterpart of gridlock on the town's narrow Main Street. Other Northeast towns celebrate their shellfishing traditions with their own versions of Oysterfest; examples include the oyster festivals in Milford, and Norwalk, Connecticut, the Pemiquid Oyster Festival in Damariscotta, Maine, the Scallop Festival in Bourne, Massachusetts, and the Taste of Providence, in Rhode Island, where oysters are featured prominently. Those who attend get the opportunity to sample the local products and learn a bit more about an industry that supports the local economy.

Harvest Methods

Before shellfish can be enjoyed and celebrated, they must be harvested from the farms on which they are grown. The methods used by growers to harvest shellfish vary with the location and depth of the stock. In deep waters, heavy cages and nets must be lifted onto a boat, often with the aid of hydraulic devices. If oysters are planted on the bottom, they must be harvested by dredging (also called "dragging" by some watermen), a method that will have an impact on bottom conditions. Technically, dredging and dragging are different operations; a drag is pulled along the bottom and skims the sediment, while a dredge has teeth and can dig a few inches into the bottom, depending on the length of the teeth. In many intertidal "grants," bagged or caged shellfish, predominantly oysters, are accessible during low tide. Bags can be opened on site or brought to shore so that oysters can be examined, culled, separated, and harvested.

In Massachusetts and Connecticut, where significant stocks of wild oysters still grow, the legal size for *Crassostrea virginica,* whether it is cultured or wild, is 3 inches in height (see fig. 4.2). In Maine, no legal size limit has been set for *C. virginica* but *Ostrea edulis* must not pass through a 3-inch gage. In Rhode Island, farmed oysters must be at least 2½ inches but many growers try to get them to 3 inches to compete with other markets.

Quahogs must be grown to legal size, at least 1 inch wide, but then can be separated further according to size for littlenecks, cherrystones, and chowders. The methods for harvest of quahogs from aquaculture farms has changed very little since the days the first rakes were designed. For the recreational clammer who is digging seeded quahogs, a tined rake or basket rake is used to find the clams and then remove them from the sediment. This activity is often referred to as "scratching" due to the sound the fork makes as it moves over the quahog's shell. Experienced clammers can read the flats to find signs of quahogs and thus eliminate random searching. When I went clamming with a pro, she found at least one quahog, often

FIG. 15.1 Quahoging with a bullrake in Pleasant Bay, Massachusetts. *Photo from the H. K. Cummings Collection; reprinted with permission of Snow Library, Orleans, Massachusetts.*

more, in each scratch by examining the characteristics of the mud and reading the tell-tale signs of the quahogs beneath, including waste products and siphon holes.

For commercial harvest of quahogs, a grower often will use a bullrake. The handle can vary in length depending on whether it will be employed in the intertidal or used from a boat in deep water (fig. 15.1). The harvester uses an alternating pull and relax motion (the reason it's also called a "jerk" rake) to pull the bullrake through the sediment with its attached basket. During the relax mode, the rake digs deeper into the sediment. The rake is taken out of the sediment with a combined twisting and flipping motion so that all the clams remain in the basket, a workout that rivals any alternative upper-body exercise. Harvested quahogs are separated according to size by hand or with a mechanical device that separates littleneck, cherrystones, and chowders while culling the undersized specimens that are destined to be returned to the "grant" until they reach legal size. In special cases, a grower may be able to obtain a permit for the harvest and sale of smaller quahogs to fill the demand by restaurants for clams in traditional Italian-style preparations of *linguini con vongole* (linguini with clams).

Softshell clams also must be raked, a down and dirty job (fig. 15.2). A short rake with long teeth, sometimes called a hoe or fork, is used and care must be taken in turning over clumps of mud, so that the shells are not damaged when raking. Harvesters also must be careful not to break smaller softshell clams, less than the legal size of 2 inches long, which are often found in the sediment above the legal specimens because of their shorter siphon length. Farmed mussels and scallops are in deep water and are harvested directly from nets and bags used for grow-out. Scallops must display a well-defined growth ring, demonstrating that they have reached the age of maturity, before they can be harvested from the wild (see fig. 4.2). This restriction may not be placed on farmed scallops, but farmed scallops must reach a certain size before they are marketable.

In all cases, recreational shellfishermen and commercial wild harvesters have a catch limit that may vary by state and town. In Massachusetts, the noncommercial shellfish limit is a 10-quart wire basketful per week per permit holder. Daily and seasonal restrictions often are imposed and shellfishing activities are limited to specific locations that are managed by the local shellfish department. Harvesters should always familiarize themselves with the local rules and regulations associated with their permit. Similar catch limits do not apply for growers, who often harvest from their "grants" to fill specific orders.

FIG. 15.2 Softshell clam harvesting in Provincetown, Massachusetts, circa late 1800s or early 1900s. *Photo: Courtesy of the National Park Service, Cape Cod National Seashore*

FIG. 15.3 Grower attaching harvester tags to oysters. *Photo: Barbara Brennessel*

Health Issues

Gone are the days when shellfishermen would sell their catch from the back of pickup trucks parked along the roadways. In the 1970s, on Long Island, the local "clam man" would have a crudely painted wooden sign near his truck advertising clams for sale at $1.00 a dozen, while fresh, shucked bay scallops went for the seemingly exorbitant price of $2.99 per pound. Aside from the current increase in price for fresh shellfish, retail practices certainly have changed.

To ensure the health of cultured shellfish, growing waters must be clean and safe, so there is considerable oversight and monitoring of water quality. However, shellfish that come from pristine growing waters may become unfit for consumption at any point between the shellfish farm and the dinner plate, including during shipping, handling, and food preparation. To further protect consumers, oversight does not stop when shellfish leave the water; states require that every batch of shellfish is tagged before leaving the farm (fig. 15.3). Harvest tags contain information about the type and amount of shellfish, the grower's name and license number, and the date and the location from which the shellfish were harvested. This tag will stay with the product when it reaches the wholesaler and must be kept on file for 90 days. The wholesaler also tags the product so that there will be a traceable path if health problems surface after shellfish are sold to a restaurant or consumer.

Handling large amounts of shellfish to fill orders from multiple clients can pose additional health problems. When a dozen Wellfleet aquaculturists banded together to form the Wellfleet Shellfish Company, their facility to collect, sort, and ship quahogs and oysters had to be approved as a Hazard Analysis and Critical Control Point (HACCP) facility. The formation of this shellfish co-op allowed the group to pool their products, meet orders from large wholesalers, and assure product on a year-round basis. In order to accomplish this while ensuring the safety of their shellfish, the facility has to meet federal food-handling specifications and be inspected regularly. From the outside, one would never suspect the high-tech nature of the plant. An extensive plumbing system circulates a supply of clean Wellfleet water, sterilized by passage through a device that delivers ultraviolet radiation. The temperature of the water is strictly maintained and oxygenated. Shellfish are kept in this wet-storage facility for a brief, approved amount of time as they are readied for shipment in refrigerated trucks. This HCCAP plant, a far cry from the neighborhood "clam man, " is an example of the technological advances and safeguards that have been put into place to keep our shellfish safe while extending their shelf life as they move from farm to table.

How to Shuck a Shellfish

In the heyday of the oyster industry in the 1800s, speed in bivalve shucking was a much-valued skill, because oysters, clams, and scallops traditionally were sold as shucked meats, with the shuckers' pay dependent on the number of animals processed. In the past, piles of shell outside the door signaled the shucking work of men, women, and children residing within a home (fig. 15.4). Shucking for volume is a tedious job that often was performed by immigrants such as Cape Verdians who were recruited to New Bedford shucking plants, where they lined up in rows on shucking tables, tossing meats into barrels while discarding shells in heaps on the floor (fig. 15.5). A proficient shucker might open five to eight hundred oysters per hour!

There is an art to shucking. Different techniques and tools are used depending on the type of bivalve. When opening oysters for the half-shell

FIG. 15.4 Shucking house on Cape Cod, circa late 1800s or early 1900s. *Photo: Courtesy of the National Park Service, Cape Cod National Seashore*

FIG. 15.5 Cape Verdeans at shucking plant in New Bedford, Massachusetts, circa late 1800s or early 1900s. *Photo: Courtesy of the National Park Service, Cape Cod National Seashore*

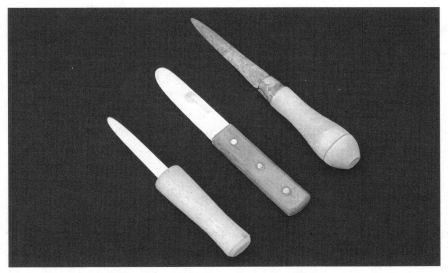

FIG. 15.6 Shucking knives. The middle blade is used to open quahogs; the other two blades are used to shuck oysters. *Photo: Nina Picariello*

consumer, a sharp, narrow, sturdy blade is employed, while clam knives tend to have a wider, flatter, blunt blade (fig. 15.6). When shucking an oyster, it's best to rinse it off first with cold water. Even experienced shuckers will use a glove or at the very least a thick towel or potholder to hold the bivalve and protect their hands. The oyster is held, cup-side (deeper side) down with the valve facing the shucker. The oyster knife is slipped into the hinge area with a slight back-and-forth motion and then a twist to pop the hinge while trying not to lose any of the oyster liquor. The meat is scraped gently from the top shell and loosened from the bottom shell with the knife. As is, or with a squirt of lemon, the oyster is ready for consumption, and usually is swallowed whole, although some prefer to chew a bit before swallowing. A fresh oyster is tasty on its own or with a bit of lemon, but oyster bars have developed the custom to offer a variety of sauces, horseradish and Tabasco to flavor the shucked oysters. Oyster bars also offer small, round or square, white, buttery crackers, known as oyster crackers. Oyster crackers were first produced in 1847 by Adam Exton, an English immigrant to New Jersey, who baked them for addition to oyster stews, in which they held up well and didn't become soggy. These crackers probably are served at oyster bars to counter the effects of the alcoholic beverages that often are ordered with a plate of oysters.

Clams have two adductor muscles holding the valves together, so a different shucking technique is used. The clam should be held in the hollow of one's hand. The clam knife is inserted between the valves and worked across the insertion point until one of the adductors is severed. The knife is twisted slightly until the shell pops open and the remains of the adductors can be cut. Some folks put their clams in the freezer for at least 15 minutes to weaken the adductors and thus make it easier to open the valves. When clams are being used for cooking purposes, rather than for the raw bar, some friends of mine rely on the powers of the microwave oven and give the clams a quick zap to make it easier to open them. It is also possible to purchase a guillotine-like device, looking a bit like an office papercutter, in which a blade can be lowered directly between the valves to open a clam.

The adductor muscle of the scallop is its most valuable feature, so care must be taken that it is not damaged during shucking. The scallop hinge should be nestled in one's palm. A strong, thin knife, such as a butter knife, is inserted between the valves and given a slight twist. The muscle is cut from the inside of the top shell and the rest of the body parts (the viscera) are pulled out and discarded. The muscle can then be cut out from the bottom shell. No matter the type of bivalve or the method of shucking, repetition and practice are the key to good shucking technique. What may be second nature for the shellfish grower or professional shucker is often a challenge for the novice.

Nutritive Value of Bivalves

Bivalve shellfish are low in fat and high in protein, which makes them a healthy food choice, according to current U.S. dietary standards. The fat content of clams and scallops (bay scallops and sea scallops) is less than 15 percent, while mussels and oysters contain 20 to 30 percent fat. Bivalve shellfish have a high proportion of good fat, that is, the polyunsaturated variety, as well as omega-3 fatty acids compared to beef, chicken, and pork, which have undetectable levels of this beneficial fat. Mussels, clams, and oysters contain two or three times as much omega-3 fatty acid as bay and sea scallops. With respect to cholesterol, bivalve molluscs contain less than 50 mg per 100 g of edible meat, compared to about 70 mg per 100 g of beef. Bivalve molluscs also have a relatively high content of the minerals iron and zinc, as well as the vitamin cobalamine (B-12), which is found only in foods of animal origin. So unless your clams, oysters, and scallops are deep fried or your culinary preparation contains unhealthy ingredients that compromise the food value, bivalve molluscs are good food items that can be part of a wholesome diet.

Bivalve Quality

In the Northeast, shellfish growers plan the operation of their aquaculture farms to optimize the production of quality oysters, most of which will be eaten raw. Restaurants and chefs are very demanding, especially when they have a choice of product from different growers. Therefore, appearance is a major factor in determining the selling price of each oyster to the wholesale market.

When I helped my neighbor Paul pick oysters to fill an order from his "grant," I was given some basic instructions. To assure the quality of Paul's oysters for the buyer, first and most importantly, the oysters had to be legal size for harvest. In Massachusetts, the specimen must measure 3 inches in height, because it is assumed that an oyster this size is mature and has had a chance to produce gametes, an important consideration in an area in which natural sets can still occur. Because it sometimes is difficult to distinguish farm-raised from wild oysters by appearance, this size requirement ensures that wild oysters have been given an opportunity to mature and reproduce, and to contribute to replenishing the natural population. Several types of gauges have been designed to facilitate rapid assessment of size, one of which is a simple metal ring with an inner diameter that defines the legal size. If the oyster falls through the ring, it is too small and cannot be harvested.

The second culling hallmark was shell appearance: A clean shell without pitting, parasites, and discoloration contributes to the value of the oyster. The third criteria I was asked to use in selecting market oysters was shape; I was asked to cull "bananas" or "rabbit ears," oysters with narrow or elongated shells that are not desired at raw bars. Only the rounder, wider oysters would be selected, because they hold a good-sized meat. The final criterion was a deep cup, a feature that indicates plenty of room to hold the briny juices or elixir when the oyster is shucked.

Appearance is not the only factor that indicates a quality oyster; taste is also a key component. Shakespeare apparently appreciated oysters, held them in high esteem, and considered them a prize. He writes in *The Merry Wives of Windsor:*

> Why, when the world is mine oyster,
> Which I with sword will conquer.

Although popular slurp-offs test how many oysters can be consumed in a given time, a more discerning culinary contest at Wellfleet Oysterfest centers around an activity that tests whether contestants can identify the point of origin of an oyster. Similar to areas used for grape cultivation, in which the natural environment and growing conditions influence the flavor of wine (terroir), analogous conditions heavily factor into the taste of oysters. In circles of oyster aficionados, I have heard the term "merroir" used to describe the unique characteristics of the water in which oysters are grown. Subtle nuances of the briny taste of an oyster are imparted by the salinity, phytoplankton variety, tidal flow, and other factors in the growing waters. (The same environmental and growing factors come into play when quahogs are grown to supply the raw "littleneck" markets.) Oyster aficionados swear that they can tell the origin of an oyster by its distinctive taste. And I'm not simply talking about East Coast versus West Coast or Chesapeake versus Louisiana; the argument can be made for different tastes within a relatively small area. During Wellfleet's Oysterfest, oyster lovers try to distinguish among the oysters from different Wellfleet oyster farms by taste alone. I doubt whether many oyster lovers have taste buds that exhibit this level of discrimination, but this contest provides an example of how the growing environment has the potential to affect taste.

A pamphlet produced by the Ocean State Aquaculture Association describes the taste of oysters from six different Rhode Island farms: "Crisp, briny flavor with slightly sweet finish"; "Fine, salty flavor"; "Salty oyster with rich flavor"; "Taste of salt air"; "Salty start, smokey finish and awesome aftertaste"; "Taste as if plucked from salted spring water."

The complex taste of New England oysters is evident from an oyster article in *Yankee Magazine* in which the taste of oysters from farms in Maine, Massachusetts, Rhode Island, and Connecticut were compared using adjectives and descriptors such as fruity, berry-like finish, light and clean brininess, lingering ocean essence, sweet and slightly nutty, silky smooth meat, creamy, nut-like, sweet seaweed flavor, coppery finish (Copps 2006).

Shucked shellfish don't move or show signs of life, except perhaps by a slight wiggle when they are squirted with lemon juice, so most individuals who eat raw shellfish don't give a thought to the fact that they are swallowing live animals. Some folks are appalled by this concept. In *Through the Looking Glass,* Lewis Carroll was aware that there might be oyster sympathizers in his audience. In staged performances of the work, he inserted extra verses after the story of "The Walrus and the Carpenter," in which the duo have tricked young oysters into accompanying them on a walk before they set to eating them. In the additional verses, the ghosts of the oysters return for revenge and sing while dancing on top of the walrus:

> O, woeful, weeping Walrus, your tears are all a sham!
> You're greedier for Oysters than children are for jam.
> You like to have an Oyster to give the meal a zest—
> Excuse me, wicked Walrus, for stomping on your chest!
> For stomping on your chest!
> For stomping on your chest!
> Excuse me, wicked Walrus, for stomping on your chest!
> (Carroll [1871] 1960, 187)

In addition to being served raw, oysters and clams can be used in many cooked recipes. Oyster recipes tend to be regional in nature but a few recipes transcend geography. For example, oyster stew and oyster dressing for Thanksgiving turkey are standards in many areas. Two oyster recipes popularized in the South are worthy of mention. I first sampled an oyster po-boy in New Orleans and still crave the taste. Po-boys are sandwiches on crispy French bread, similar to subs in New England and heros in New York. Oysters Rockefeller, so named to indicate the richness of the dish, was created by a chef at Antoine's, one of New Orleans' most famous French Quarter restaurants (see Appendix for recipes).

Perhaps the earliest cooked shellfish preparation, employed by the early New England natives, corresponds to what we now call a clambake. The Wampanoags chose rocks of uniform size and shape and mounded them in a shallow circular pit. Dry wood was placed on top of the rocks and a fire was started. When the rocks were glowing hot, the wood was removed and the clams were added to the pile. The whole "bake" was covered with rock-weed (Mills and Breen 2001), which was kept steaming by juices liberated from the clams, supplemented with the addition of measured amounts of seawater. The clambake was adapted by European settlers and was employed as a venue for social events (fig. 15.7). In today's version, clams (usually steamers and/or littlenecks), corn on the cob, potatoes, mussels, chicken, linguica (spicy sausage), and lobsters may be on the menu. Due to the difficulty of finding suitable sites and prohibitions against fires on the beach, the same ingredients sometimes are steamed on a rack within a large pot with water on the bottom, thus converting the clam bake to the modern clam boil.

Softshell clams are used most often as "steamers," quickly steamed with a little water until they open, then served with the strained clam broth and

FIG. 15.7 Cape Cod clambake at Billingsgate Island: *top* tending the fire pit; *bottom* the guests, who arrived by boat, enjoy the clambake, circa early 1900s. *Photos courtesy of the National Park Service, Cape Cod National Seashore*

some butter. The steamer is removed from the shell by hand, held by the neck, dipped in the broth to remove residual sand, then dipped in butter and eaten.

Chowder-size quahogs, ocean quahogs, and surf clams are used to make clam chowders: the thick, creamy New England version, the thinner, tomato-based Manhattan chowder, or the Rhode Island recipe that turns out a clear chowder (see the appendix for chowder recipes). Manhattan clam chowder, a tomato-based soup, originally called Coney Island Clam Chowder, was popularized in the 1930s. It is believed to have originated with Italian immigrants in New York and Portuguese immigrants in Rhode Island. New Englanders consider it a major offense to add tomatoes to clam chowder and a popular myth holds that it is against the law to put tomatoes in clam chowder in Maine and Massachusetts. Indeed, a Maine legislator tried to promote a bill in 1939 that would make it illegal to put tomatoes in clam chowder, but the bill was never passed. Although it's clear where New Englanders stand in the milk versus tomatoes debate, a check at the Massachusetts Law Library did not turn up any legislation on the issue.

Eating Out

Many people who are unlikely to harvest shellfish on their own or who don't have proficiency with shucking are more comfortable eating shellfish at a raw bar or restaurant. The Union Oyster House in Boston has been serving raw oysters since 1826, making it the oldest restaurant in Boston and the oldest restaurant in the entire United States in continuous service. Originally named the Atwood and Bacon Oyster House, after its original owners, it was, and still is, famous for its semi-circular raw bar that was frequented by the movers and shakers of Boston society, including Daniel Webster, who drank "a tall tumbler of brandy and water with each half dozen oysters, seldom having less than six plates" (Union Oyster House n.d.). The Union Oyster House was a favorite gathering spot for Boston's first family, the Kennedys; a booth has been dedicated to the memory of John Fitzgerald Kennedy and the establishment is still patronized by local politicians. Union Oyster House only serves very fresh and very local oysters and clams with the point of origin of littlenecks and cherrystones identified as Cape Cod.

Perhaps the most famous raw bar in the Northeast is in the unlikely locale of a train station: the Grand Central Oyster Bar, opened in 1913 in Grand Central Terminal in the heart of New York City. Five thousand oysters are shucked and sold per day, with two to three dozen different types of oysters represented on the menu. For the discerning raw bar patron, oysters are listed by popular name and point of origin: Beaver Tails, Moonstones, Narragansetts, Watch Hills, and Ninigrets from Rhode Island; Belons from Nova Scotia; Bluepoints, Widow's Holes, East Ends, Great

South Bays, and Pipes Coves from Long Island: Bogue's Bays and Chincoteagues from Virginia: Caraquets from New Brunswick; Conway Cups and Malpeques from Prince Edward Island; Cuttyhunks, Wellfleets, Whitecaps, Martha's Vineyards, and Plymouth Rocks from Massachusetts; Eagle Rocks, Otters Coves, Qulcenes, Kunamotos, Yaquinas, and Wescott Bay Flats from Washington State; Glidden Points and Pemaquids from Maine; and Stonington Selects from Connecticut. Most of these offerings, notably those from the northeastern and middle Atlantic, are specimens of the eastern oyster, *Crassostrea virginica,* but some are European flat oysters, *Ostrea edulis,* Pacific oysters, and other varieties. For those who prefer cooked oysters, roasted, sautéed, and smoked preparations are available.

The most likely place where one would sample bivalve shellfish in the Northeast, with family in tow, is in a typical "clam shack," some of which have become very popular and somewhat more sophisticated eating establishments. Although simple clam shacks dot the East Coast, more commonly, the shacks have become seafood restaurants that cater to the whole family and individuals with diverse tastes. Moby Dick's (for a whale of a meal), in Wellfleet, has morphed from a simple hot dog stand to a very popular seafood restaurant where hungry tourists form long lines to place their orders before they can sit down to alfresco dining in the screened porch. Owners Todd and Mignon Barry use only local Wellfeet oysters and littlenecks, assuring their customers of the freshest product for their raw bar. Clam chowder, always a year-round favorite in New England, is made from surf clams (*Spisula solidissima*). Mignon Barry maintains that *Spisula* doesn't get as tough as quahogs in chowder preparations. Moby Dick's, like many other restaurants, uses the same basic, commercially available surf clam stock, but adds its own secret ingredients to give their chowder a distinctive flavor.

Whether one eats bivalve shellfish raw or cooked, at home or outside the home, these morsels are quite delicious. Rather than a staple food item, bivalves are considered to be delicacies, and as such, are not found routinely in small grocery stores and restaurants. Although there is not a high demand from the average family or individual, a lucrative market has evolved in the Northeast for a "boutique" product, with restaurants typically charging $2.00 per raw oyster. With the incentive to supply this specialty market, the shellfish culture industry continues to have economic importance.

Chapter 16

The Current State of Aquaculture in the Northeast

At the 2006 Northeast Aquaculture Conference and Exposition (NACE) held in Mystic, Connecticut, many presentations began with a quote by the late, great, marine biologist and filmmaker, Jacques Cousteau. Cousteau said, "We must plant the sea and herd its animals using the sea as farmers instead of hunters." Today, Cousteau's vision is being realized. Many species of fish, including shelled varieties, are being grown in aquaculture facilities to satisfy the demand for this high-protein product. Shellfish aquaculture is a true farming operation and a billion-dollar business in the Unites States. Only the states of Washington and Florida have more shellfish farms than Massachusetts (USDA 2005). Of 980 shellfish farms nationwide, 223 are located in New York and New England, where significant increases in the number of farms has occurred since the last United States Department of Aquaculture (USDA) Census in 1998. The breakdown and comparison to 1998 shows the contribution of the various New England States and New York to the industry (table 16.1).

Shellfish and Shellfish Aquaculture Organizations

A survey of aquaculture and shellfish organizations dredges up a plethora of acronyms that cover most of the alphabet. Many organizations have been formed to support the shellfish industry, conduct research related to aquaculture and shellfish restoration, and facilitate interactions among government agencies, growers, and scientists. Congress has established five regional aquaculture centers to address issues and opportunities that vary with geographic location. The Regional Aquaculture Centers are administered by the USDA under their Cooperative State Research, Education and Extension Service (CSREES). The aquaculture extension services in each state provide the same categories of support for shellfish farmers as farming extension services do for traditional agricultural businesses encompassing a wide array of crops. The Northeastern Regional Aquaculture Center

Table 16.1

United States Shellfish Farms in the Northeast

State	Number of shellfish farms		2005 sales (in thousands)
	1998	2005	
Maine	15	32	$2,861
New Hampshire	0	2	ND
Massachusetts	94	138	$6,157
Rhode Island	2	11	ND
Connecticut	15	27	ND
New York	2	13	ND
Total U.S.	535	980	$203,183

ND: Not disclosed.

Source: USDA, *Census of Aquaculture,* 2005.

(NRAC) recently moved its offices from the University of Massachusetts at Dartmouth to the University of Maryland at College Park. According to the NRAC website, http://wwwnrac.umd.edu/aboutnrac/mission.html, the mission of the organization is to serve as a "forum for the advancement and dissemination of science and technology needed by Northeastern aquaculture producers and support industries." NRAC promotes research that helps the aquaculture industry identify and solve problems and assists with large projects that smaller state or local agencies alone would not be able to handle. NRAC also supports educational initiatives that have a direct effect on the industry.

In addition to NRAC, many national and international committees and organizations have been formed to address various aspects of aquaculture. Some organizations focus on the commercial and economic aspects of the industry, some on health and sanitation issues, and others on scientific research and restoration of shellfish areas. Melbourne Carriker, a renowned malacologist, chronicles the formation of various U.S. organizations related to the shellfish and aquaculture industries in his book, *Taming of the Oyster* (Carriker 2004). As a result of the decline in the oyster industry in the late 1800s and early 1900s, an organization, the Oyster Growers and Dealers Association (OGDA), was formed to deal with issues related to growing, marketing, and sanitation of oysters. OGDA was a trade association whose mission was to turn around the industry. Its political operating arm was OINA, the Oyster Institute of North America.

The National Shellfish Association (NSA) evolved in 1930 from its origins as the National Association of Shellfish Commissioners (NASC) in 1909 and its reorganization in 1915 as the National Association of Fisheries Commissioners (NAFC). NSA promotes the interaction among shellfishermen, government agencies and scientists and publishes the premier journal in the field of shellfish research, *The Journal of Shellfish Research* (*JSR*), which originally was entitled *Proceeding of the National Shellfisheries Association* (*PNSA*).

The NSA met regularly with the growers' groups OGDA-OINA in joint conventions for many years at the beginning of the twentieth century. In 1970, OINA became the Shellfish Institute of North America (SINA) to reflect the broadening of its mission to include other molluscs and also crustaceans. SINA continued to meet on a regular basis with NSA, and in 1996, morphed into the Molluscan Shellfish Institute (MSI). The Interstate Shellfish Sanitation Commission (ISSC) emerged in 1982 as a partnership between the Food and Drug Administration, state fish-regulation agencies, and shellfish industry interests. ISSC took up human health issues that were emerging as a result of shellfish grown in polluted waters, and subsumed the roles of OGDA-SINA in 1996. Growers associations have flourished on both coasts of the United States to represent industry concerns. In the Northeast, the East Coast Shellfish Growers Association (ECSGA) has represented the industry at local and national levels.

The various shellfish aquaculture associations and agencies meet regularly in regional, national, and international venues to share ideas, address problems, and guide the direction for research. The meetings are an interesting mix of university scientists, representatives from town, state, and federal agencies, and shellfish growers. The overarching, common goals for these diverse groups are the development and support of shellfish aquaculture as an economical, healthy, sustainable, and environmentally sound method of providing shellfish for human consumption. Another goal, not necessarily shared by all who are involved in shellfish aquaculture, but of interest to some will be much more difficult to achieve: restoration of the historic shellfish beds in our coastal areas. Bivalve restoration efforts will be described in the next chapter.

Chapter 17

Aquaculture and the Environment

Finfish aquaculture has been criticized by environmentalists for its potential degradation of ecosystems. Aside from being a monoculture, finfish aquaculture, which involves raising animals such as trout, talapia, and salmon in crowded pens, often necessitates the use of antibiotics to prevent widescale spread of infectious diseases among the animals. These chemicals, which leach into the environment, put selection pressure on microbes and foster the development of microbial resistance. Large, concentrated exogenous food is added to the fish pens and large amounts of nitrogenous waste products are released into the water, contributing to pollution and algal blooms. Scientists and environmentalists have been calling for technological changes in the industry to render finfish farming less destructive to marine and freshwater ecosystems.

In contrast, cultured shellfish do not require the addition of food or treatment with antibiotics. Some waste products are released into the environment in the form of feces, pseudofeces, and ammonia; however, these products also are released from natural oyster reefs and shellfish beds. I do not make the claim that shellfish farming is without environmental impact. Shellfish farms undoubtedly have an effect on the bottomlands as a result of the deployment of gear and in particular, dredging equipment that accompanies the larger farming operations. But, in comparison to current finfish farming practices, shellfish farming appears more environmentally benign.

Destruction of eelgrass beds is a prime example of a negative consequence of some types of shellfish aquaculture. Eelgrass stabilizes marine sediments, protects shorelines from erosion, and provides habitat and nursery for other organisms, including fish, shellfish, and crustaceans. Some aquaculturists argue that shellfish gear creates habitat and attracts reef-dwelling organisms. However, gear provides an artificial habitat and although it may attract many organisms, the assemblage may differ from that found on natural reefs or bottoms.

Many unresolved questions still exist regarding the changes in benthic communities, as well as other components of the ecological food web, as a result of dredging and deployment of shellfish aquaculture gear. For example, some scientists have suggested possible changes in the distribution of sediment-dwelling invertebrates that provide much-needed food for migrating shore-birds and even lack of access of migrating birds to certain feeding grounds. Another environmental aspect of shellfish aquaculture that has been targeted by environmentalists is the degradation of sensitive coastal areas as a result of off-road vehicle traffic across dunes and tidal flats so that growers can have access to their "grants."

Some upland property owners and folks who appreciate a pristine vista might argue that shellfish farms and gear contribute to a type of visual pollution, but so do mobile-phone towers, power lines, and other structures that we can't seem to live or function without. In moderation, and suitably located, deployed shellfish farming gear might even be considered a natural feature of the maritime economy, similar to piers, docks, and marinas.

Although the visual impact of shellfish aquaculture can be debated and the possible detriment to bottomlands and coastal ecology can be posited, there is one, undeniably positive value in shellfish aquaculture: filtration of coastal waters and removal of potentially problematic levels of phytoplankton. For example, while Long Island's brown tides of the 1980s and 1990s are believed to have caused the dramatic declines in Peconic Bay shellfish, some marine experts hypothesize that shellfish declines preceding the brown tide resulted in loss of filtration capacity and contributed to the overgrowth of *Aureococcus,* the brown tide culprit. The best-known example of the filtration capability of shellfish is given in Chesapeake Bay statistics, which calculate that prior to modern declines, Chesapeake oysters were responsible for the massive filtration of the entire bay (see chapter 2). Although no one has done the calculation for the Northeast, our shellfish farms are taking up some of the missing filtration and plankton-removal roles that formerly were accomplished by wild shellfish populations.

Shellfish Restoration

Although aquaculture may satisfy the demands for market shellfish destined for restaurants and home use, positive and necessary ecological contributions of these filter feeders are not being met by artificial cultures. Native shellfish beds perform vital roles in the environment, labeled as "ecosystem functions" by biologists. Why are natural shellfish beds important? It can't be overemphasized that shellfish are instrumental in filtering phytoplankton and nitrates, thus purifying the water, improving water quality, and allowing more light to penetrate and reach underwater grasses that provide important habitat for many species. In addition, oyster reefs serve as strategic barriers that mitigate the effects of storm surges, slow down erosion, and serve as refuge and spawning habitat for other aquatic species. Furthermore, bivalves are important components of the food chain, consumed by invertebrates, birds, mammals, and even other molluscs. Bivalves return nutrients to the environment in the form of waste, utilized and recycled by bacteria, and shell, which stabilizes the bottom and provides setting sites for larvae. Cultured shellfish, which are raised in cages, bags, and nets, and then removed from the water, cannot fulfill all of these important ecosystem roles. It is ironic that in order to have ecologically important wild stocks of oysters, scallops, and quahogs, it is becoming more and more necessary to depend on shellfish hatcheries and artificial cultivation techniques. Thus it is only with the help of farming that we will be able to transform shellfish "crops" into natural "inhabitants" of marine ecosystems.

As natural stocks of bivalve shellfish have become depleted in many parts of their range, individuals and organizations are making concerted efforts to restore some of the populations, with the overall goal of reaching historic numbers. Large-scale oyster restoration projects are underway on the West Coast, where the State of Washington is striving to restore the native Olympia oysters, *Ostrea conchaphila,* which has largely disappeared while many different oyster species introduced from Asia and Europe are grown in large commercial shellfish farms. The native Olympia is a small oyster, 2 inches in height, which was relegated to second-class status after the importation of larger oyster species for aquaculture. The Olympia has been declining gradually under pressure from shoreline development and habitat loss (Gordon, Blantan, and Nosho 2001). Recently, the cultural value of the Olympia oyster has become appreciated by West Coast native tribes. Furthermore, it has been discovered that the native Olympia fulfills ecosystem functions that are not being met by the new imports. For example, the smaller Olympia forms more densely packed reefs that provide a unique habitat, compared to the large, fast-growing, cultivated Pacific oyster, *Crassostrea gigas,* imported from Japan in the early 1900s. These native oysters support different communities of organisms compared to those found in clustered shells of larger oysters.

To understand some of the problems inherent in large-scale shellfish restoration projects, lessons can be learned from considering Chesapeake

Bay, an area in which the historic oyster population has declined dramatically due to pollution, overharvest, and disease. The depletion of oyster reefs has led to a loss of the ecosystem function of the oysters, as well as a decline in commercial harvest. This dire situation has prompted heroic attempts to restore the oyster resources of the bay; aquaculture plays a key role in these efforts. Scientists are breeding disease-resistant oysters that may survive Dermo and MSX (see chapter 13). Cultch is being deposited in attempts to attract spat and rebuild historic reefs.

Perhaps the most controversial proposal for the restoration of oysters in Chesapeake Bay was an attempt in 2005 to put *Crassostrea virginica* on the list of endangered species. Dieter Busch of Ecosystems Initiatives Advisory Services in Crownsville, Maryland, had a radical idea. He petitioned Congress under the Endangered Species Act of 1973 to list the Eastern oyster as threatened or endangered due to overharvest, disease, and especially, water quality. Such a listing would provide enormous protection of native oysters and dramatically limit their harvest, not only in the Chesapeake, but throughout the nation. It was acknowledged that the Chesapeake watershed needs to be cleaned up, but the Congressional hearing concluded that this type of listing was not the best way to accomplish the goal (Congressional Hearing Report 2005). Moreover, according to the criteria for listing species in the endangered category, there was a lack of scientific justification for such a classification for *C. virginica*. Needless to say, this proposal did not sit well with the oyster industry nor with baymen who make a living at oystering.

A more recent Chesapeake Bay initiative for oyster restoration has the potential to have an impact on oysters throughout the Northeast. Maryland and Virginia have made a controversial proposal to bring a nonnative oyster to Chesapeake Bay, similar to the import of European and Japanese oysters to Washington State. The Asian oyster, *Crassostrea ariakensis,* is native to China, and studies have shown that it is much more resistant to oyster diseases such as Dermo and MSX than native oysters. Proposals have been advanced to use the Asian oyster for commercial aquaculture as well as for restoration projects. Despite the potential of *C. ariakensis* to fix the problems in Chesapeake Bay, there is also potential for this introduction to be a disaster. Very little is known about the potential ecological and economic impact of such an introduction. Following the recommendations of the International Council for the Exploration of the Sea in their "Code of Practice on the Introductions and Transfers of Marine Organisms," the U.S. Congress in 2003 mandated the preparation of an Environmental Impact Statement (EIS) to examine possible benefits and possible risks of the introduction. A multi-agency task force is involved with federal input via the Army Corp of Engineers, state agencies from Maryland and Virginia, and the cooperation of the U.S. Environmental Protection Agency (EPA), NOAA, and FWS (Fish and Wildlife Service).

This potential introduction of *C. ariakensis* is not a trivial issue. Congressional hearings (Congressional Hearing Report 108-67), 2005 mandated further study and the annual budget for research on *C. ariakensis* and its biological and ecological roles, overseen by NOAA's Chesapeake Bay Office, is $2 million. Scientists are studying such aspects as the genetics of the species to determine if this oyster is truly a single species, the susceptibility of *C. ariakensis* to parasites and pathogens, possible human health risks from consumption of diseased *C. ariakensis,* interactions of *C. ariakensis* with native oysters, and ecosytem services of *C. ariakensis* compared to *C. virginica.* (Does it form reefs? What is its filtration rate?) Most importantly for aquaculture, scientists are providing information that will be used to decide whether the future of aquaculture in Chesapeake Bay will focus on native oysters or nonnative oysters.

How can the introduction of a nonnative oyster to Chesapeake Bay affect oysters in the Northeast? Quite simply, by virtue of the ability of oysters to release gametes, there is the possibility that nonnative oysters may hybridize with *C. virginica.* No physical barriers exist to keep the gametes or their progeny within Chesapeake Bay. The resulting growth characteristics, geographical range, ecology, taste, and other aspects of a new type of oyster cannot be predicted. In an extreme situation, if nonnative oysters reproduce and can outcompete native oysters for food and habitat, native oysters might disappear entirely. To address this very important issue regarding introduction of nonnative species, oyster biologists have manipulated *C. ariakensis* so that it can't reproduce. Oysters can be rendered mutant by treatment with a chemical agent that interferes with cell division and produces tetraploids, cells with four sets of chromosomes instead of two (diploid). These tetraploids are bred in hatcheries with normal diploid oysters and the resulting progeny are sterile triploids (having three sets of chromosomes). Theoretically, triploid *C. ariakensis* would not have the ability to breed and populate nor hybridize with *C. virginica,* so this is considered a type of biosecurity. An added benefit of triploid oysters is that they grow faster; the energy that would be used for reproduction is put into growth of the meat and shell. The only drawback of this technology is that biologists cannot guarantee that 100 percent of the progeny is triploid; a few "normal" oysters may result from the selective breeding procedures. If that is the case, there would be a potential for the spread of this nonnative oyster within the Chesapeake and beyond.

A multitude of shellfish restoration projects are proposed or underway in every state in the Northeast, with native scallops, quahogs, and oysters the focus of most of the attention. Not surprisingly, most attempts to restore native shellfish center on the use of hatchery-raised bivalves that are grown to planting size by employment of aquaculture techniques. I wish I could report on attempts that have resulted in astounding success stories, but these types of projects proceed with baby steps, sometimes with setbacks.

Sandy Macfarlane, working in Orleans, Henry Lind in Eastham, and Rick Karney on Martha's Vineyaerd are among the Cape Cod pioneers in shellfish restocking programs (Macfarlane 2002). Sometimes called renourishment or enhancement, early Cape Cod restocking programs initially were designed and supported to boost the commercial and economic aspects of the industry. These initiatives, which ramped up in the 1970s, have evolved into natural resource projects that focus not only on the replenishment of shellfish stocks, but also on key environmental issues such as pollution and coastal development.

Shellfish restoration is a complicated endeavor involving not only adding to the number of shellfish, but also restoring, enhancing, and maintaining proper habitat for them to grow and reproduce. These are often massive projects that involve many partners: conservation groups, university researchers, private companies, state agencies, cities, and towns. A prime example of such a partnership is the Bluepoints Bottomlands Project, whose partners include the Bluepoints Oyster Company, The Nature Conservancy, the town of Brookhaven, Long Island, the New York State Department of Environmental Conservation, the State University of New York Marine Science Division at Southampton, Long Island, and Cornell Cooperative Extension. This project, which involved the acquisition of 13,000 acres of bottomland, encompasses 30 percent of Great South Bay. The holdings can be traced to a purchase in 1694 by William Smith that eventually became part of the Bluepoints Oyster Company, one of the largest oyster companies on the East Coast. Most of the bottom, valued at about $2 million, was donated by Bluepoints; the balance was purchased by The Nature Conservancy for restoration, education, and research with the eventual goal of supporting a sustainable harvest. According to Nature Conservancy figures, the harvest of quahogs from Great South Bay in 1976 accounted for 50 percent of the nation's total catch, but by 2003, less than

10,000 bushels were harvested (Brumbaugh et al. 2006) This project involved the creation of hard clam spawner sanctuaries—i.e. areas closed to harvest—where millions of hatchery-raised quahogs were stocked. It is hoped that these no-take zones will provide broodstock that will help to repopulate Great South Bay with clams. The Nature Conservancy also has developed spawner sanctuaries for clams and scallops in Peconic Bay.

Another example of a complex restoration effort from Long Island, New York, points to efforts to clean up the water to protect the resource and help to remediate environmental and economic damage. The Peconic Estuary Program takes a multifaceted approach to habitat restoration that can ultimately benefit the shellfishing industry and shellfish aquaculture. With farmland surrounding the watershed, habitat restoration involves initiatives that address farm waste issues and runoff, ranging from pesticides to manure generated from duck farms.

In Connecticut, federal, state, and local agencies have been working on similar approaches in shellfish restoration with a focus on estuarine health. Some projects have been more successful than others and it's not surprising to learn that the more successful projects occurred in areas with cleaner waters. The Oyster River Shellfish Restoration Project in Old Sayville, Connecticut, is a community-based project that involves partnerships among federal, state, municipal, and private agencies. It provides an example of the cooperation and organization needed to conceptualize and implement these large-scale ecosystems projects.

One of the questions that can be asked about shellfish restoration is: What broodstock is being used for a particular project? With modern hatchery techniques and genetic selection, one can select for fast growers, disease-tolerant specimens, or even shell attributes. By selecting for certain traits, broodstock can become somewhat inbred and unknown deleterious genes may be carried along with the genetic traits that are deemed desirable. Inbred lines have the potential to hybridize with wild stocks, which might be beneficial to the population or, depending on the outcome, could be very problematic. In areas with a residual wild shellfish population, some inherent problems with the use of selected lines of shellfish may not be apparent immediately but may surface in years to come.

Shellfish Gardening

Some restoration programs that rely on hatchery-reared bivalves depend on shorefront property owners to provide nursery habitat, thus allowing the shellfish to grow to a suitable size before they are used for reseeding areas that are targeted for restoration. Hence, shellfish gardens are blooming near restoration areas. I first learned about shellfish gardening at a meeting of the Friends of Pleasant Bay when Kim Tetrault gave an overview of a community-based shellfish restoration program, fittingly named SPAT (Southold Program in Aquaculture Training). This SPAT is a different organization than Wellfleet's SPAT but the acronym fits the mission of both groups. The SPAT program in Peconic Bay is overseen by Cornell Cooperative Extension of Suffolk County, New York, at their Marine Environmental Learning Center. The year-round curriculum involves lectures and projects to educate community members to plant, culture, and monitor the growth of shellfish on their own waterfront property or in the SPAT facility. The goal of the program is to restore the shellfish to Peconic Bay by increasing environmental awareness and encouraging local citizens to be responsible for environmental stewardship. The SPAT volunteers, who include carpenters, plumbers, electricians, physicians, undertakers, retirees, and families, get to keep a portion of the oysters: The rest are used for restoration. Tetrault hopes that eventually small-scale oyster farming will support a boutique oyster industry on the North Fork of Long Island. To this end, SPAT teams with local vineyards to sponsor wine and oyster tasting events as venues to introduce folks to wines produced from locally grown grapes and to local, farm-raised oysters.

I went to see the SPAT operation for myself and arrived just in time for the cowbell that announced a morning coffee break for volunteers. Tetrault gave me a tour of the facility and I met some of the SPAT members who are involved in all phases of the program: operating and maintaining the hatchery and nursery upwellers, grow out, monitoring water quality, shellfish research, public relations, fundraising, youth education, giving presentations, and other important functions. This noncommercial venture uses several grow-out sites, closed to harvest, to produce scallops and oysters to repopulate the Bay. A new shipment of lantern nets was being unpacked and awaited the receipt of bay scallops, destined for grow-out at a deepwater site on the very eastern end of Long Island (see fig. 10.4). When mature, the scallops would be seeded into Peconic Bay, where, until the mid-twentieth century, a major scallop fishery existed. The project was designed so that the hatchery-reared scallops would help to restore the population of scallops by serving as broodstock and also would contribute to the improvement of water quality in the bay. In 2004, New York State invested $1.75 million in this project

SPAT members have been major contributors of time and labor to various projects. This energetic group of volunteers constructed some of the outbuildings at the Marine Environmental Center and, on the day of my visit, were putting together a sleek dory as a raffle prize for one of their fundraisers. SPAT is an example of the benefit of channeling the nonstop energy, camaraderie, experience, and ideas of local citizens to make a long-lasting difference in the environment.

Another oyster garden is in bloom in Rhode Island, under the blue thumb of Dr. Dale Leavitt, Aquaculture Extension agent for Rhode Island. The Roger Williams University hatchery provides the seed oysters, while local citizens provide the shorefront or dock space for grow out. The oysters are then shipped to and deposited in some of the smaller embayments of Narragansett Bay where Leavitt can monitor their progress.

Whether it is a municipal, state, or federal project, the shellfish industry and environmental groups agree about the importance of restoring shellfish to areas in which they were once abundant. These efforts will continue to require the infusion of funds to explore a number of strategies and to determine which approach is optimal in individual areas.

Chapter 18

The Future of Shellfish Aquaculture

We don't need a crystal ball to understand why harvest of native shellfish throughout the world has not proved to be a sustainable endeavor. As Pollan states in *The Omnivore's Dilema,* "Fishing is the last economically important hunter-gatherer food chain, though even this foraging economy is rapidly giving rise to aquaculture, for the same reasons hunting wild game succumbed to raising livestock. It is depressing though not at all difficult to imagine our grandchildren living in a world in which fishing for a living is history" (Pollan, 2006, p. 280).

The same principle applies for shellfishing. Although individual shellfish are fecund, producing millions of gametes which result in legions of larval creatures, the survival of individual organisms has a very low degree of probability. When we add habitat degradation, pollution, overharvest, and even global warming to the mix, it's difficult to understand why there are any wild shellfish at all. Harvest of wild shellfish is simply unsustainable: there are too many shellfishermen and too much efficient, high-tech harvest equipment, contributing to the complete depletion of native stocks. This has happened time and again in every region of the world.

In contrast to the wild shellfishery, the future of aquaculture in the northeast currently looks more promising, There is demand for high quality, locally grown, shellfish. The technique of shellfish farming allows the product to be harvested to order, thus guaranteeing freshness. The animals are grown without pesticides or chemicals and are supplied without additives. Shellfish are essentially an organic food in terms of the manner in which they are farmed and harvested.

Expansion and Improvement

Does the future of the shellfish industry reside in the expansion of shellfish aquaculture? In most areas of the Northeast, aquaculture is now the only method that is bringing shellfish to the table. Some locales, like the state of

Rhode Island, are promoting shellfish aquaculture as a growth industry. However, at other sites, there is little room for growth or an overwhelming number of restrictions. There are also questions about the "carrying capacity" of certain areas and their ability to support sufficient growth to make it profitable. Presumably, if there are too many shellfish farms in a limited area, the phytoplankton levels will not support realistic rates of growth of the product.

There are some serious impediments to shellfish aquaculture expansion, the most important of which is water quality; many sites are simply unsuitable for growing shellfish for today's market. In Connecticut, where the slogan "As fresh as it gets" is promoted by the Connecticut Seafood Council, other factors that stand in the way of shellfish aquaculture have been identified. A survey conducted in 2003, by the Connecticut SeaGrant Extension Program to determine the needs of the industry, collected information from producers, producer associations, regulators, state departments of agriculture and environmental protection and extension agents by questionnaire, interviews and focus groups (Duff et al., 2003). Although the findings reflect the current state of shellfish aquaculture in Connecticut, most of the issues are typical of other northeast states. Communication and relationship building were identified as areas needing improvement, with a call to improve relationships among shellfish culturists and the public and an imperative to streamline the permitting process. A large gap was identified between public awareness of shellfish aquaculture and public acceptance of the industry. The issue of user conflict still looms large (Getchis, 2005). Some actionable items that were identified as a result of the Connecticut survey are also universally applicable. There should be more research to determine the ecological consequences of shellfish aquaculture, in terms of the species cultured and the methods and equipment used (Getchis, 2005).

Documents such as Best Management Practices (BMPs) for the Shellfish Culture Industry in Southeastern Massachusetts, a voluntary set of guidelines, describe cultivation techniques and practices that assist the growers, while stressing low negative impact on the environment. The practices were developed by the growers themselves in collaboration with SEMAC (South-Eastern Massachusetts Aquaculture Center) and not only have the potential to improve the operation from the economic standpoint of the grower, but also address environmental impacts and user conflicts. In addition to BMPs, individual towns in Massachusetts are encouraged to adopt their own Shellfish Management Plans to address local issues and to shape the future direction of the industry.

The issues surrounding user conflicts will not be easily resolved and competition for finite tracts of coastline will likely increase. Many of the shorefront areas in which shellfish aquaculture is currently undertaken or have been identified as potential sites for expansion are also areas that depend heavily on tourism. The same individuals and families who patronize seafood restaurants or buy fresh, local shellfish are attracted to these locations because of opportunities for boating, swimming and other forms of shoreline recreation. If these opportunities decline as a result of aquaculture expansion, tourists and vacationers will travel elsewhere and towns will lose a major source of revenue.

There are rising concerns that the federal government might put some restrictions or changes on the further development and expansion of shellfish aquaculture operations. In an initiative to simplify permitting and encourage individuals and companies to invest in various types of sustainable fish farming projects, a task force was organized by researchers from Woods Hole Oceanographic Institute, with funding from the PEW Charitable Trust and the Lenfast Foundation. While recognizing the importance of the aquaculture industry and the need for growth to compensate for declining wild fish stocks, the task force also focused on environmental responsibility. In addition to suggesting that fish farming operations, in general, might be better sited offshore rather than near shore, there was a movement to prohibit farming operations in areas with SAV (submerged aquatic vegetation) as part of the nationwide permitting process. This initiative is based on a serious environmental issue: the dramatic decline in SAV, most notably eelgrass. Dragging the bottom and shading the bottom with gear have the potential to further decimate eelgrass or prevent it from becoming established. The environmental issues raised by the task force led to the publication of a draft document in 2007 on Nationwide Permit D (U.S. Army Corps of Engineers 2007) which has already been subjected to a public comment period in which shellfish industry organizations and growers have weighed in. Needless to say, the industry, especially larger shellfish farming companies, is concerned about the potential impact of the task force recommendations to the future of the business.

Fiscal Concerns for Growers

With economic uncertainty surrounding the industry, it might be difficult to attract a new generation of shellfish growers. Shellfish diseases and predation continue to take their toll while storms can decimate a grower's crop and gear. Losses are common and most shellfish aquaculturists feel lucky if they can grow half their seed into market size animals. Vandalism and poaching occur with regular frequency. When the cost of running a shellfish aquaculture operation is taken into consideration, including boats, crews, managers, electricity, fuel etc., the balance sheet may run in the red. There may be added costs if the shellfish has to be depurated or stored for any period of time or if it must be shipped via refrigerated truck. Sometimes, these costs are those of the wholesaler, but will be reflected in the price paid to the grower. There remains a great deal of uncertainty regarding the actual financial picture emerging from shellfish growing operations. It is generally acknowledged that shellfish harvest statistics and shellfish farm income is grossly underreported. For example, the Massachusetts Department of Marine Fisheries assumes that these numbers might be from 40 to 60 percent higher than reported (Massachusetts Office of Coastal Zone Management, 1995).

Economic help may be available for the quahog grower in the form of crop insurance which covers some losses due to storms, disease and other factors. The U.S. Department of Agriculture Risk Management Agency issues federal cop insurance subsidies, which assume that 40% of a quahog crop will be lost to natural factors and assumes that older clams, near market size, are worth more than seed. Premiums, on a sliding scale, depend on the extent of coverage desired by the grower. Clam losses are covered for decreases in salinity, ice floe damage to gear, freezing, hurricanes, storm surges, tidal waves and disease. Thus far, there are no policies which cover similar losses of farmed oysters.

Global Climate Change

The long term future of shellfish aquaculture may also be tied to an environmental factor that has reached the forefront of world-wide concerns—global climate change. The predicted climate changes may have far-reaching impacts on the shellfish industry with potentially negative effects coming from a variety of directions. Increases in temperature and greenhouse gas levels will produce a combined threat to shellfish. Warming of the ocean waters could lead to a redistribution of bivalve species. In a surprising 2005 observation, blue mussels were reported to have been detected in the high Arctic where they have not been present for about a thousand years. Surf clams, ocean quahogs and other cold water species might be expected to shift further northward in their distribution and warm water species may even find the coastal northeast too warm for comfort and move farther off-shore. Global warming may eventually foster the rapid evolution of shellfish species that tolerate the changing waterscape but might not be so good to eat. Problems such as red tides could be more frequent and more widespread.

Warmer ocean temperatures will also increase the incidence of seafood-associated illness caused by bacteria such as *Vibrio*. An example of such an occurrence affected passengers aboard an Alaska cruise ship in 2004. An outbreak of gastroenteritis with symptoms such as diarrhea was associated with *Vibrio parahaemolyticus,* serotype O6:K18, isolated from the sick passengers as well as samples of raw oysters that they consumed. This pathogen is usually found in warmer southern waters and not at the latitude of the Alaskan oyster farm from which the oysters originated. In this case, the oyster disease vectors were harvested when average daily water temperatures exceeded 15°C (59°F), the presumptive threshold temperature for risk of gastroenteritis from consuming raw oysters infected with *Vibrio parahaemolyticus*. Environmental monitoring produced evidence that there was a trend of rising water temperature in the Gulf of Alaska, the location of the shellfish farm that supplied the oysters to the cruise ship (McLaughlin et al., 2005).

In addition to the potential impact on human consumers, warming oceans may bring new predators and invasive species which will be able to spread north from warmer, southern waters, and, as polar ice melts, more freshwater will flow into the ocean and reduce its salinity. Global warming will also affect the circulation of ocean waters and resulting currents that supply food for bivalve molluscs.

Rising global carbon dioxide (CO_2) is not limited to the atmosphere but is also absorbed by the marine environment where it can change ocean chemistry. It has been estimated that about 30% of the CO_2 that enters the atmosphere from the burning of fossil fuel finds it way into the oceans. When CO_2 dissolves in water, some of it forms carbonic acid (H_2CO_3; in weak solution it is used in carbonated soft drinks), which, in turn, partially dissociates into hydrogen and bicarbonate ions. Thus, when CO_2 is found in marine waters it forms a complex mixture of gaseous CO_2, dissolved CO_2, carbonic acid and bicarbonate. In general, the hydrogen ions acidify water and thus lower the pH. Marine environments tend to have slightly basic pH values, 8.0 to 8.3. If the pH is decreased, even slightly, the effect on marine life may be profound. There is some evidence, from controlled laboratory studies, that lowered pH will inhibit shellfish, and other organisms such as coral, from incorporating calcium carbonate into their structure. The hydrogen ions can also combine with free carbonate ions (CO_3), forming bicarbonate (HCO_3) and thus diminishing the availability of carbonate itself, which is the structural building block incorporated by many organisms, including bivalve shellfish, to build shell. Furthermore, species of phytoplankton that use calcium carbonate in their structure will not be able to grow. Thus, the entire marine food chain will be seriously affected. There are even predictions that some marine organisms will actually lose mineralized calcium carbonate and begin to disintegrate (Orr et al., 2005). Mussels are believed to be the most at-risk bivalve, but oysters, clams and scallops are also expected to suffer.

An additional threat to shellfish farming is the inevitability of rising sea level. Tidal flats are the ideal location for shellfish farms in many northeast areas because they allow easy access to the growing organisms, thus allowing for frequent monitoring and maintenance. As sea level rises, these intertidal locations will be obliterated and transformed into sub-tidal, or deepwater areas that may abut existing coastal development or be part of private property. Shellfish growing methods and equipment, used in the northeast may have to change as growing areas become submerged.

To respond to rising sea level, off shore, open ocean aquaculture, used for finish farming and mussel culture may become the method of choice for oysters and scallops. To address the potential expansion of aquaculture into federal waters in an environmentally responsible manner, the Marine Fisheries Advisory Committee requested that NOAA develop a 10 Year Plan for Marine Aquaculture. The Plan, finalized in 2007, proposes regional ecosystem management initiatives that have several desired outcomes, including not only a viable and prosperous aquaculture industry but also the restoration of wild stocks. The plan proposes to "restore and expand coastal shellfish aquaculture in areas with appropriate carrying capacity and establish and expand the use of offshore methods for production of oysters, mussels, abalone and (possibly) scallops" (NOAA 2007).

Ebb or Flow of Shellfish Aquaculture?

What's on the horizon for bivalve shellfish aquaculture? With the advantage of hindsight and, in the current climate of environmental awareness, the buzzword for any type of food production is sustainability. Is shellfish farming sustainable? Can shellfish continue to be farmed without degrading the environment or depleting marine resources? If I were forced to answer that question, I would answer, "It depends . . . " At this point in time, it is imperative that there are concerted efforts to restore natural stocks of shellfish, which have plummeted, partially as a result of unsustainable fishing practices. Shellfish farming takes pressure off the limited wild stocks. Furthermore, farmed shellfish do not need an exogenous source of food or result in the nitrogen loading of sensitive coastal waters. In addition, small scale shellfish farms rely on manual labor and do not use gas guzzling mechanical equipment which can contribute to global CO_2 increases. Technology and practices are improving to protect other species that share the flats with farmed shellfish and shellfish farmers are becoming some of the most vocal advocates for clean water and coastal restoration projects. If conducted in a responsible manner, using best management practices and with appropriate federal, state and municipal oversight, shellfish aquaculture as practiced in the northeast, need not be an enemy of the environment and may even go a long way to help in the restoration of historic shellfish beds.

The future of shellfish farming in the northeast is thus dependent on many factors including those that are somewhat controllable or can be manipulated, such as supply and demand, public perception, environmental impacts and local and government policies, as well as factors which are less controllable such as the increasing prevalence of shellfish diseases and most importantly, global climate change. These factors will mold and shape the direction of the industry in the near term and in the years to come.

It is not only critical that aquaculture be encouraged and supported so that we can supply shellfish for their nutritive and culinary value; more importantly, from an environmental perspective, bivalve shellfish will continue to be cultured and grown for critical restoration initiatives, to prevent depletion from their natural habitats and to ensure their continued biofiltration function in coastal ecosystems. We must keep our waters clean and free of pollution in order to grow and consume shellfish, and the shellfish will continue to pump and filter our water and keep it clear and free of phytoplankton blooms that can cause hypoxia and degrade coastal ecosystems. Thus, it is important that shellfish farms remain as prints on the tides and continue to be part of the coastal symbiosis between humans and bivalves.

RECIPES

Coquille St. Jacques

This is a good way to use previously frozen bay scallops from China. I do not recommend this preparation for fresh bay scallops because they are too sweet and tasty to be smothered with a creamy sauce.

1 lb bay scallops

1 bay leaf

½ lemon, sliced

salt and pepper

½ cup dry white wine

½ cup water

3 tbsp. butter

1 cup sliced mushrooms

2 tablespoons minced onion or shallots

2 tablespoons flour

½ cup cream or milk, depending on preference

½ cup fine dry bread crumbs

⅓ cup grated Parmesan cheese or Italian cheese mixture (Romano, Asago, Fontina, and Parmesan available in any supermarket).

Place scallops in pan with onion, bay leaf, lemon slices, wine and water. Bring to a simmer and cook about 3–5 minutes until scallops are firm. Remove the scallops with a slotted spoon and transfer them to a lightly buttered pie plate or baking dish; some folks use individual ramekins, but I prefer to make one family-style plate. Reserve the broth but remove lemon slices and bay leaf. Next, sauté onion (or shallots) and mushrooms in butter, then over a low flame, blend in the flour and reserved scallop liquid, stirring constantly until thickened. Ladle sauce over the scallops, sprinkle breadcrumbs and cheese on top, and place pan under broiler for 10–15 minutes. Watch carefully to be sure that the top browns slightly but does not burn. Serves 4.

Moules Mariniere

This is an easily prepared dish that works well as a meal or appetizer.

3 to 4 lbs of mussels. Farm raised mussels are usually sold in a ready to cook state. If mussels have beards, trim the beards and gently scrub the shells,

2 tablespoons of butter or extra virgin olive oil

1 tablespoon minced garlic

1 minced onion

1 lemon

1 bunch fresh parsley, chopped

1 cup dry white wine

salt and pepper to taste

Sauté the garlic and onions in the butter or olive oil. When onions become translucent, add the wine, parsley, and the juice of the lemon. Bring the mixture to a boil; then add the mussels, lower to a simmer, and cover. Simmer 5 to 10 minutes, stirring or shaking the pot occasionally to keep the mussels distributed. When mussels have opened, remove them with slotted spoon and place them in a serving dish. Discard any unopened mussels. Pour the cooking liquid over the mussels and serve with a crusty bread. Serves from 3 to 4, or more if served as an appetizer.

Picariello Linguini and Clams

You will need about 8 to 12 large quahogs for each pound of linguini. Shuck the clams over a bowl so that you can collect the juice. Pour the juice into a glass or measuring cup and let settle. Decant the juice and pour it through a kitchen strainer. This step removes all the sand and shell bits.

Chop the clam meats into small pieces and saute for a few minutes with several chopped cloves of garlic. Add the clam juice to the pot with clams and garlic. Add about one part water or white wine to two parts clam juice (otherwise, the sauce gets too salty). Bring almost to a boil, lower heat, and add fresh parley and red and black pepper to taste. Pour over cooked linguine. Serves 4.

Scungilli Marinara

Here is a recipe for a spicy dish prepared with whelks, called *scungilli* in Italian. This dish, which is often served as an appetizer, may make some converts.

½ lbs scungilli, thinly sliced	2 cups canned tomatoes
3 tablespoons olive oil	2 tablespoons tomato sauce
2 cloves garlic	½ teaspoon salt
1 minced onion	½ teaspoon each oregano and basil
1 minced celery stalk	¼ teaspoon hot red pepper seeds

Boil scungilli for 15 minutes and drain. Pull out the meats and slice into thin sections. Place scungilli with oil, garlic, onion, and celery in a skillet and sauté until light brown. Add tomatoes, tomato sauce, and salt, and slowly cook until scungilli is tender. Add oregano, basil, and hot pepper, and cook 5 more minutes. Serve with crusty bread. (Adapted from Ada Boni, *The Talisman Italian Cookbook*.)

Rhode Island Clam Chowder

Note: the quahog juice is salty, so it's best to hold off on adding salt until the chowder is ready to be served.

3 lbs red potatoes, diced	2 cups quahog juice
¾ lb salt pork or bacon	4 teaspoons fresh thyme
1 red onion, diced	salt and pepper to taste
2 cups chopped quahogs	

In a large pot, cover potatoes with two inches of water and bring to a boil, then lower to a simmer. In a separate pan, heat the salt pork or bacon, and add onions. Sauté until onions are translucent but not browned. Add pork and onions to potatoes, followed by quahogs, quahog juice, and thyme. Simmer until potatoes are cooked through but not mushy. Add salt and pepper to taste. Serves 15. (Adapted from Gabriella True, "Culinary discovery: The Tastes of Rhode Island," on Senior Women Web, www.senior women.com/hs/articles/true/articlestrueRhode.html.)

New England Clam Chowder

This preparation is an amalgam of several recipes and one that I have used with considerable success. It is similar to the Rhode Island version above except that the onion can be white or yellow and milk or cream is included. Follow the Rhode Island Chowder recipe above. When potatoes are just about cooked, add 2 cups of heated skim milk, low-fat milk, whole milk, or cream (depending on your preference). Remove the salt pork or bacon, and season with salt, pepper, and chopped parsley. Some folks like to add a dab of butter to each bowl while serving. This recipe produces a chowder of somewhat soupy consistency. To make a thick chowder, take a table-spoon of flour in a bowl, add some of the hot clam juices from your soup pot, and stir until the lumps have disappeared and you have a smooth white paste. Return the paste to the soup pot and cook for a few minutes longer. The chowder will thicken considerably.

Manhattan Clam Chowder

To make the Manhattan variety of clam chowder, add a diced celery stalk, and a cup or two of peeled, diced tomatoes to the basic Rhode Island recipe. The resulting concoction can be flavored with a bit of Worcester-shire sauce. An easier way to sample the combination of clams with toma-toes is to purchase commercially available mixtures of tomato juice and clam juice, aptly named "clamato" juice.

Oyster Stew for Two

I adapted this quick recipe for oyster stew from several versions prepared by oyster growers who prepare it with their legal-sized, farmed oysters that don't pass the raw bar beauty criteria.

1 onion, diced	1 potato, cooked and diced
1 dozen oysters; meats and juices separated	1 to 2 cups of warmed milk or cream (this can be heated in a microwave oven)

Sauté the onion in some butter or olive oil. Add the oystermeat and con-tinue to sauté for a few minutes, then add the juices and simmer for a few minutes to heat up the mixture. Add potatoes and warmed milk or cream, and heat through but do not boil. Season with salt, pepper, and fresh chopped parsley.

Oyster Po-Boy

My variation serves two or more, depending on the size of the loaf and the number of oysters. Drain shucked oysters and coat with corn-meal flour, seasoned with flaked red pepper. Fry oysters in hot oil and drain on a paper towel. Split and toast a French bread loaf and slather it with melted butter. Distribute fried oysters on one half of the loaf. Squeeze some lemon juice over the oysters and cover with the other half of the loaf. Eat immediately while oysters are warm and bread is crisp.

Oysters Rockefeller

The original version contained spinach, but now there are many variations. The following basic recipe has been successful in my kitchen.

After shucking, place the oysters in the half-shell on their cupped side on a baking sheet. A bed of rock salt can be used to keep the oysters in place. Top the oysters with a mixture containing the following ingredients per dozen oysters:

6 tablespoons olive oil or butter	½ teaspoon Pernod, anisette, or anisone (the original recipes called for absinthe, but any anise-flavored liqueur can be used)
6 tablespoons raw chopped spinach	
3 tablespoons minced onion	
3 tablespoons chopped fresh parsley	pinch salt
6 tablespoons bread crumbs	Tabasco to taste

Sauté the above ingredients for 5 to 10 minutes, until onions are clear. Cool, then place on top of the oysters. Broil oysters until just brown. Serve with wedges of lemon.

BIBLIOGRAPHY

Abbott, R. Tucker. 1968. *A Guide to Field Identification: Seashells of North America.* Edited by Herbert S. Zim. New York: Golden Press, Western Publishing Company.

Abraham, B. J., and P. L. Dillon. 1986. "Species Profiles: Life Histories and Environmental Requirements of Coastal Fishes and Invertebrates (Mid-Atlantic)—Softshell Clams." *U.S. Fish and Wildlife Service Biological Report* 82, no. 11.68, U.S. Army Corps of Engineers. TREL-82-4.

Audemard, C., L. M. Ragone Calvo, K. T. Paynter, K. S. Reece, and E. M. Burreson. 2006. "Real-Time PCR Investigation of Parasite Ecology: *In Situ* Determination of Oyster Parasite *Perkinsus Marinus* Transmission Dynamics in Lower Chesapeake Bay." *Parasitology* 132: 827–42.

Baptist, G., D. Meritt, and D. Webster. 1993. "Growing Microalgae to Feed Bivalve Larvae." *NRAC Fact Sheet.* North Dartmouth, Mass: Northeast Regional Aquaculture Center, U.S. Department of Agriculture.

Beauchamp, William M. 1901. *Wampum and Shell Articles Used by the New York Indians.* Bulletin of the New York State Museum, vol. 8. Albany: University of the State of New York.

Belding, David L. 1910. *Report Upon the Scallop Fisheries of Massachusetts.* Department of Marine Fisheries, Commonwealth of Massachusetts.

———. 1912. *Report Upon the Quahog and Oyster Fisheries of Massachusetts.* Division of Marine Fisheries, Commonwealth of Massachusetts.

———. [1930] 2004. *The Quahog Fisheries of Massachusetts.* Department of Marine Fisheries, Commonwealth of Massachusetts. Reprinted in *The Works of David L. Belding, M.D.* Cape Cod Cooperative Extension.

———. 1915. *Report Upon the Soft-Shell Clam Fisheries of Massachusetts.* Department of Marine Fisheries and Game, Commonwealth of Massachusetts.

———. [1931] 2004. The Scallop Fishery of Massachusetts. Department of Marine

Fisheries, Commonwealth of Massachusetts, *Marine Fisheries Series* no. 2. Reprinted in *The Works of David L. Belding, M.D.* Cape Cod Cooperative Extension.

Beninger, P. G., and M. LePennec. 2006. "Structure and Function in Scallops." In *Scallops: Biology, Ecology and Aquaculture,* edited by S. E. Shumway and G. J. Parsons. *Developments in Aquaculture and Fisheries Science,* vol. 35, 2nd ed. New York: Elsevier Amsterdam Press.

Boni, Ada. 1974. *The Talisman Italian Cook Book.* Translated by Matilde Pei. Special edition printed for the Ronzoni Macaroni Co. New York: Crown Publishers.

Bower, S. M. 2006a. "Synopsis of Infectious Diseases and Parasites of Commercially Exploited Shellfish: *Haplosporidium nelsoni* (MSX) of Oysters." Fisheries and Oceans Canada, http://www-sci.pac.dfo-mpo.gc.ca/shelldis/pages/hapneloy_e.htm.

———. 2006b. "Synopsis of Infectious Diseases and Parasites of Commercially Exploited Shellfish: *Perkinsus marinus*" ("Dermo" Disease) of Oysters. Fisheries and Oceans Canada, http://www-sci.pac.dfo-mpo-gc.ca/shelldis/pages/pmdoy_e.htm.

Brait, Susan. 1990. *Chesapeake Gold: Man and Oyster on the Bay.* Lexington: University Press of Kentucky.

Brazee, Shanna L., and Emily Carrington. 2006. "Interspecific Comparison of the Mechanical Properties of Mussel Myssus." *Biological Bulletin* 11: 263–74.

Brooks, William K. [1891] 1996. *The Oyster.* Baltimore: Johns Hopkins University Press.

Brumbaugh, R. D., M. W. Beck, L. D. Coen, L. Craig, and P. Hicks. 2006. *A Practitioner's Guide to the Design and Monitoring of Shellfish Restoration Projects: An Ecosystem Approach.* Arlington, Va.: The Nature Conservancy.

Calvo, Lisa M. Ragone, and Eugene M. Burreson. 2002. *QPX Susceptibility in Hard Clams Varies with Geographic Origin of Brood Stock,* Virginia Sea Grant Marine Resource Advisory Program No. 74, Vsg-02-18. College of William & Mary, Virginia Institute of Marine Science.

Calvo, L. M., G. W. Calvo, and E. M. Burreson. 2003. "Dual Disease Resistance in a Selectively Bred Eastern Oyster, *Crassostrea virginica,* Strain Tested in Chesapeake Bay. *Aquaculture* 220: 69–87.

Carriker, Melbourne R. 2001. "Functional Morphology and Behavior of Shelled Veligers and Early Juveniles." In *Biology of the Hard Clam,* edited by J. N. Kraeuter, and M. Castagna. Amsterdam: Elsevier Science.

———. 2004. *Taming of the Oyster: A History of Evolving Shellfisheries and the National Shellfisheries Association.* Hanover, Penn.: The Sheridan Press.

Carriker, M. R., and Patrick M. Gaffney. 1996. "A Catalogue of Selected Species of Living Oysters (*Ostreacea*) of the World." In *The Eastern Oyster* "Crassostrea virginica," edited by Victor S. Kennedy, Roger I. E. Newell, and Albert F. Eble, 1–18. College Park, Md.: Maryland Sea Grant College.

Carroll, Lewis. [1871] 1960. *Through the Looking Glass.* In *The Annotated Alice.* Edited by Martin Gardner. New York: W.W. Norton & Co.

Castagna, Michael. 2001. "Aquaculture of the Hard Clam, *Mercenaria mercenaria.*" In *Biology of the Hard Clam,* edited by J. N. Kraeuter, and M. Castagna, 675–79. Amsterdam: Elsevier Science.

Castagna, Michael, and John N. Kraeuter. 1981. *Manual for Growing the Hard Clam* "Mercenaria." Gloucester Point, Va: Virginia Institute of Marine Science.

Chesapeake Bay Foundation. n.d. "Oyster Fact Sheet." www.clof.org/site/PageServer ?pagename=resources_facts_oysters.

Churchill, E. P., Jr. 1920. *The Oyster and the Oyster Industry of the Atlantic and Gulf Coasts.* Washington, D.C.: Government Printing Office.

Clark, Eleanor. 1959. *The Oyster of Locmariaquer.* New York: Pantheon Books.

Clark, N. T. 1931. "The Wampum Belt Collection of the New York State Museum." *New York State Museum Bulletin* 288:85–121.

Congressional Hearing Report. 2003. "Efforts to Introduce Non-Native Oyster Species to the Chesapeake Bay and the National Research Council's Report Titled 'Non-Native Oysters in the Chesapeake Bay'" no. 108-67, October 14. U.S. Government Printing Office, http://www.access.gpo.gov/congress/house.

Congressional Hearing Report. 2005. "Listing of the Eastern Oyster under the Endangered Species Act" No. 109-24. U.S. Government Printing Office, http://www.gpoaccess .gov/congress/index.html.

Copps, A. B. 2006. *Our World Is an Oyster.* Dublin, N.H.: Yankee Publishing.

Corbett, Scott. 1955. *Cape Cod's Way: An Informal History of Cape Cod.* Binghamton, N.Y.: Vail Ballou Press.

Cox, Ian, ed. 1957. *The Scallop: Studies of a Shell and Its Influences on Humankind.* London: Shell Transport and Trading Company, Limited.

Coyne, Kathryn J., Xiao-Xia Qin, and Herbert J. Waite. 1997. "Extensible Collagen in Mussel Byssus: A Natural Block Copolymer." *Science* 277, no. 5333: 1830–32.

Cragg, Simon M. 2006. "Development, Physiology, Behavior and Ecology of Scallop Larvae." In *Scallops: Biology, Ecology and Aquaculture,* 2nd ed. Developments in Aquaculture and Fisheries Science no. 35, edited by Sandra E. Shumway and G. Parsons. Amsterdam: Elsevier.

Dahl, S., M. Perrigault, and B. Allam. "Tolerance of Different Hard Clam Stocks to Various Isolates of Quahog Parasite Unknown (QPX)." 2006. International Conference on Shellfish Restoration Abstracts no. 9.

Day, Michael J., Dean E. Franklin, and Bonnie L. Brown. 2000. "Use of Competitive PCR to Detect and Quantify *Haplospordium nelsoni* Infection (MSX Disease) in the Eastern Oyster (*Crassostrea virginica*)." *Biotechnology* (March): 456–65.

Dickens, Charles. 1843 "A Christmas Carol" Electronic Text Center. University of Virginia Library. http://etext.virginia.edu/toc/modeng/public/DicChri.html.

Duff, J. A., T. S. Getchis, and P. Hoagland. 2003. "A Review of Legal and Policy Constraints to Aquaculture in the U.S. Northeast." *Aquaculture White Paper* no. 5, NRAC Publication no. 03-005.

Dupuy, John L., Nancy T. Windsor, and Charles E. Sutton. 1977. "Manual for Design and Operation of an Oyster Seed Hatchery for the American Oyster *Crassostrea virginica.*" Gloucester Point: Virginia Institute of Marine Science/SEA Grant.

Eble, Albert F., and Robert Scro. 1996. "General Anatomy." In *The Eastern Oyster* "Crassostrea virginicus," edited by Victor S. Kennedy, Roger I. E. Newell, and Albert F. Eble, 19–73. College Park, Md.: Maryland Sea Grant College.

Ewart, John W., and Susan E. Ford. 1993. "History and Impact of MSX and Dermo Diseases on Oyster Stocks in the Northeast Region." *NRAC Fact Sheet* 200-1993. North Dartmouth: University of Massachusetts-Dartmouth.

Fay, C. W., R. J. Neves, and G. B. Pardue. 1983. "Species Profiles: Life Histories and Environmental Requirements of Coastal Fishes and Invertebrates (Mid-Atlantic)— Bay Scallop." Division of Biological Services. FWS/OBS-82/11.12. U.S. Army Corps of Engineers, TR EL-82.

Fisher, M. F. K. [1941] 1988. *Consider the Oyster.* San Francisco: North Point Press.

Fitt, W. K., and S. L. Coon. 1992. "Evidence for Ammonia as a Natural Cue for Recruitment of Oyster Larvae to Oyster Beds in a Georgia Salt Marsh." *The Biological Bulletin* 182, no. 3: 401–408.

Flimlin, Gef, and Brian F. Beal. 1993. "Major Predators of Cultured Shellfish." *NRAC Bulletin* 180-1993. North Dartmouth, Mass.: Northeast Regional Aquaculture Center.

Ford, Susan E. 2001. "Pests, Parasites, Diseases, and Defense Mechanisms of the Hard Clam, *Mercenaria mercenaria.*" In *Biology of the Hard Clam,* edited by J. N. Kraeuter and M. Castagna. Amsterdam: Elsevier Science.

Ford, Susan E., and M.R. Tripp. 1996. "Diseases and Defense Mechanisms." In *The Eastern Oyster* "Crassostrea virginica," edited by Victor S. Kennedy, Roger I. E. Newell, and Albert F. Eble, 581–660. College Park, Md.: Maryland Sea Grant College.

Galpin, Virginia M. 1989. *New Haven's Oyster Industry 1638–1987.* New Haven, Conn: New Haven Colony Historical Society.

Galtsoff, Paul S. 1964. *The American Oyster:* "Crassostrea virginica gmelin." Washington, D.C.: United States Government Printing Office.

Getchis, T. S. 2005. "An Assessment of the Needs of Connecticut's Shellfish Aquaculture Industry." Connecticut Sea Grant College Program, CTSG-05-02.

———. 2006. "Clam Heaven." *Wrack Lines* 6, no. 11 (Connecticut Sea Grant): 3–6.

Getchis, T., L. Williams, and A. May. 2006. *Seed Oystering.* Connecticut Sea Grant Publication. Groton, Conn.: University of Connecticut.

Glude, John B. 1957. "Copper, a Possible Barrier to Oyster Drills." *Proceedings of the National Shellfisheries Association* 47: 73–82.

Goedkin, Michael, Brenda Morsey, Inke Sunila, and Sylvain De Guise. 2005. "Immunomodulation of *Crassostrea gigas* and *Crassostrea virginica* cellular defense mechanisms by *Perkinsus marinus.*" *Journal of Shellfish Research* 24, no. 2: 487–96.

Gordon, David G., Nancy E. Blanton, and Terry Y. Nosho. 2001. *Heaven on the Half Shell: The Story of the Northwest's Love Affair with the Oyster.* Portland, Ore.: Washington Sea Grant Program and Westwinds Press.

Gosling, Elizabeth. 1992. *The Mussel Mytilus: Ecology, Physiology, Genetics, and Culture.* Amsterdam: Elsevier.

———. 2003. Bivalve Molluscs. *Biology, Ecology and Culture.* Oxford: Blackwell Publishing.

Guillard, R. L. 1983. "Culture of Marine Invertebrates: Selected Readings." In *Culture of Phytoplankton for Feeding Marine Invertebrates,* edited by Jr. C. J. Berg. Stroudsberg, Penn.: Hutchinson Ross Publishing.

Guo, X, S.E. Ford, G. DeBrosse, R. Smolowitz, and I. Sunila. 2003. "Breeding and Evaluation of Eastern Oyster Strains Selected for MSX, Dermo and JOD Resistance." *Journal of Shellfish Research* 22: 333–34.

Hardin, Garrett. 1968. "The Tragedy of the Commons." *Science* 162, no. 3859: 1243–48.

Hart, Deborah R., and Antonie S. Chute. 2004. *Essential Fish Habitat Source Document: Sea Scallop,* "Placopecten magellanicus," *Life History and Habitat Characteristics,* 2nd ed. Woods Hole, Mass.: National Marine Fisheries Service, Woods Hole Laboratory, September.

Haskin, Harold H., and Jay D. Andrews. 1988. "Uncertainties and Speculations about the Life Cycle of the Eastern Oyster Pathogen *Haplosporidium* nelsoni (MSX)." In *Disease Processes in Marine Bivalve Molluscs,* edited by William S. Fisher. Special Publications, 18th ed. Bethesda, Md.: The American Fisheries Society.

Hedeen, Robert A. 1986. *The Oyster—the Life and Lore of the Celebrate Bivalve.* Centerville, Md.: Tidewater Publishers.

Helm, M. N., and N. Bourne. 2004. "Hatchery Culture of Bivalves: A Practical Manual." *FAO. Fisheries Technical Paper* 471. Rome: Food and Agriculture Organization of the United Nations.

Hickey, M. 2002. "Shellfish Sanitation and Management." Massachusetts Division of Marine Fisheries, http://www.mass.gov/dfwele/dmfl/programsandprojects/shelsani.htm.

Inoue, Koji, Yasuhiro Takeuchi, Daisuke Miki, and Satoshi Odo. 1995. "Mussel Adhesive Plaque Protein Gene Is a Novel Member of Epidermal Growth Factor-Like Gene Family." *Journal of Biological Chemistry* 270, no. 12: 6698–701.

Jacobsen, Larry, and James Weinberg. 2006. "Ocean Quahog." *Status of Fisheries Resources in the Northeastern United States.* NEFSC, NOAA. http://www.nefsc.noaa.gov/sas/spsyn/iv/quahog.

Jordan, L. 1998. "Money Substitutes in New Netherlands and Early New York: Wampum." University of Notre Dame, Department of Special Collections, http://www .coins.nd.edu/ColCoin/ColCoinIntros/NNWampum.html.

Kennedy, Victor S. 1996. "Biology of Larvae and Spat." In *The Eastern Oyster* "Crassostrea virginicus," edited by Victor S. Kennedy, Roger I. E. Newell, and Albert F. Eble, 371–421. College Park, Md: Maryland Sea Grant College.

Kennedy, Victor S., Roger I. E. Newell, and Albert F. Eble. 1996. *The Eastern Oyster:* "Crassostrea virginica." College Park, Md.: Maryland Sea Grant College.

Kochiss, John M. 1974. *Oystering from New York to Boston.* Middleton, Conn.: Wesleyan.

Kraeuter, John N., and Michael Castagna, eds. 2001. *Biology of the Hard Clam, Developments in Aquaculture and Fisheries Science.* Amsterdam: Elsevier.

Kurlansky, Mark. 2006. *The Big Oyster: History on the Half Shell.* New York: Ballantine Books.

Leavitt, Dale, and Tracy Crago. 1996. "The Case of the Dying Quahogs: A Scientific Mystery Unfolds." *Nor'easter* (Spring/Summer) [Magazine of the Northeast Sea Grant Programs].

Leng, Bo, Xiao-Dan Liu, and Qing-Xi Chen. 2005. "Inhibitory Effects of Anticancer Peptide from *Mercenaria* on the Bgc-823 Cells and Several Enzymes." *Federation of European Biochemical Societies* 579, no. 5: 1187–90.

Loosanoff, Victor L., and Harry C. Davis. 1963. "Rearing of Bivalve Mollusks." In *Advances in Marine Biology,* edited by F. S. Russel. New York: Academic Press.

Lucas, Jared M., Eleonaro Vaccaro, and Herbert J. Waite. 2002. "A Molecular Morphometric and Mechanical Comparison of the Structural Elements of Byssus from *Mytilus edulis* and *Mytilus galloprovincialis.*" *Journal of Experimental Biology* 205: 1807–17.

Luckenbach, Mark W., and Jake Taylor. 1999. "Oyster Gardening in Virginia." Virginia Institute of Marine Sciences, School of Marine Sciences, College of William and Mary.

Lyons, M. Maille, Roxanna Smolowitz, Christopher Dungan, and Steven B. Roberts. 2006. "Development of a Real Time Quantitative PCR Assay for the Hard Clam Pathogen Quahog Parasite Unknown (QPX)." *Diseases of Aquatic Organisms* 72: 45–52.

Lyons, M. Maille, J. Evan Ward, Roxanna Smolowitz, Kevin R. Uhlinger, and Rebecca J. Gast. 2005. "Lethal Marine Snow: Pathogen of Bivalve Mollusc Concealed in Marine Aggregates." *Limnology and Oceanography* 50, no. 6: 1983–88.

Macfarlane, Sandy. 2002. *Rowing Forward Looking Back: Shellfish and the Tides of Change at the Elbow of Cape Cod.* South Orleans, Mass.: Friends of Pleasant Bay.

MacKenzie, Clyde L., Jr. 1996. "History of Oystering in the United States and Canada, Featuring the Eight Greatest Oyster Estuaries." *Marine Fisheries Review* 53.

MacKenzie, Clyde L., Jr., Allan Morrrison, David L. Taylor, Victor G. Burrell, Jr., William S. Arnold, and Armando T. Wakida-Kusunoki. 2002. "Quahogs in Eastern North America: Part I, Biology, Ecology, and Historical Uses." *Marine Fisheries Review* 64, no. 2: 1–55.

MacKenzie, Clyde, Jr., and Robert Pikanowski. 1999. "A Decline in Starfish, *Asterias forbesi,* Abundance and a Concurrent Increase in Northern Quahog, *Mercenaria mercenaria,* Abundance and Landings in the Northeastern United States." *Marine Fisheries Review* 61, no. 2: 66–71.

Manzi, J. J., and Michael Castagna, eds. 1989. *Clam Mariculture in North America.* Developments in Aquaculture and Fisheries Science. New York: Elsevier.

Martien, Jerry. 1996. *Shell Game: A True Account of Beads and Money in North America.* San Francisco: Mercury House.

Massachusetts Office of Coastal Zone Management. 1995. "Massachusetts Aquaculture White Paper—Shellfish Bottom and Off-Bottom Culture." www.mass.gov/CZM/wpshell.htm.

Massachusetts Office of Coastal Zone Management. 2007. "Massachusetts Aquaculture White Paper—Seafood Safety/Public Health." www.mass.gov/czm/wpsfsph.htm.

Matthiessen, George C. 1992. *Perspective on Shellfisheries in Southern New England.* Essex, Conn.: The Sounds Conservancy, Inc.

McCay, Bonnie F. 1998. "Oyster Wars and the Public Trust." *Property, Law, and Ecology in New Jersey History.* Tucson: The University of Arizona Press.

McGlathery, Karen J. 2001. "Macroalgal Blooms Contribute to the Decline of Seagrass in Nutrient-Enriched Coastal Waters." *Journal of Phycology* 37, no. 4: 453–56.

McHugh, J. L. and M. W. Sumner. 1988. *Annotated Bibliography II of the Hard Clam* "Mercenaria mercenaria." NOAA Technical Report NMF 68. Seattle, Wash.: U.S. Department of Commerce, NOAA, National Marine Fisheries Service.

McLaughlin, Joseph B, Angelo DePaola, Cheryl Bopp, Karen A. Martinek, Nancy P. Naplilli, Christine G. Allison, Shelley L. Murray, Eric C. Thompson, Michele M. Bird and John Middaugh. 2005. "Outbreak of *Vibrio parahaemolyticus* Gastroenteritis Associated with Alaskan Oysters." *The New England Journal of Medicine* 353: 1463–70.

Mills, Earl, Sr., and Betty Breen. 2001. *Cape Cod Wampanoag Cookbook—Wampanoag Indian Recipes, Images and Lore.* Santa Fe, NM: Clear Light Publishers.

Naidu, K. S., and G. Robert. 2006. "Fisheries Sea Scallop, *Plactopectin magellanicus.*" In *Scallops: Biology, Ecology, and Aquaculture,* edited by Sandra E. Shumway and G. Parsons. Developments in Aquaculture and Fisheries Science no. 35. Amsterdam: Elsevier.

Newell, Roger I. E., and Christopher J. Langdon. 1996. "Mechanisms and Physiology of Larval and Adult Feeding." In *The Eastern Oyster* "Crassostrea virginica," edited by Victor S. Kennedy, Roger I. E. Newell, and Albert F. Eble, 185–229. College Park, Md.: Maryland Sea Grant College.

New York State Department of Environmental Conservation, 2005. *Environment DEC Newsletter,* June. http://www.dec.state.ny.us/website/environmentdec/2005a/raritan 50205.html.

NOAA. 2007. "NOAA 10-Year Plan for Marine Aquaculture." U.S. Department of Commerce, National Oceanographic and Atmospheric Administration. NOAA Aquaculture Program website, http://aquaculture.noaa.gov/about/tenyear.html.

Orchard, William C. 1975. *Beads and Beadwork of the American Indians.* New York: Museum of the American Indian.

Orr, James C., Victoria Fabry, Oliver Aumont, Laurent Bopp, Scott C. Doney, Richard A. Feely, Arnand Gnanadesikan, et al. 2005. "Anthropogenic Ocean Acidification over the Twenty-First Century and Its Impact on Calcifying Organisms." *Nature* 237: 681–86.

Parsons, G. Jay, and Shawn M. C. Robinson. 2006. "Sea Scallop Aquaculture in the Northwest Atlantic." In *Scallops: Biology, Ecology and Aquaculture,* ed. Sandra E. Shumway and G. Jay Parsons. Developments in Aquaculture and Fisheries Science, No. 35. Amsterdam: Elsevier.

Pechenik, Jan A. 1996. *Biology of the Invertebrates,* 4th ed. New York: McGraw Hill Publishers.

Penna, Maria-Soledad, Mazhar Kahn, and Richard A. French. 2001. "Development of a Multiplex PCR for the Detection of *Haplosporidium nelsoni, Haplosporidium costale* and *Perkinsus marinus* in the Eastern Oyster (*Crassostrea virginica, gmelin,* 1971)." *Molecular and Cellular Probes* 15: 385–90.

Peters, Esther C. 1988. "Recent Investigations on the Disseminated Sarcomas of Marine Bivalve Molluscs." *American Fisheries Society Special Publication* 18: 74–92.

Pollan, Michael. 2006. *The Omnivore's Dilemma: A Natural History of Four Meals.* New York: Penguin Press.

Rice, Michael A. 1992. *The Northern Quahog: The Biology of* "Mercenaria mercenaria." Narragansett: Rhode Island Sea Grant.

Riisgard, Hans Ulrik. 1998. "Efficiency of Particle Retention and Filtration Rate in 6 Species of Northeast American Bivalves." *Marine Ecology—Progress Series* 45: 217–23.

Rippey, Scott R. 1994. "Infectious Diseases Associated with Molluscan Shellfish Consumption." *Clinical Microbiology Reviews* 7, no. 4: 419–525.

Rivara, Gregg, Kim Tetrault, and R. Michael Patricio. 2002 . *A Low Cost Floating Upweller Shellfish Nursery System.* Cornell University Cooperative Extension of Suffolk County. http://counties.cce.cornell.edu/suffolk/MARprograms/Aquaculture/upweller factsheet1/pdf.

Russell, Spencer, Jr., Salvatore Frasca, Inke Sunila, and Richard A. French. 2004. "Application of a Multiplex PCR for the Detection of Protozoan Pathogens of the Eastern Oyster *Crassostrea virginica* in Field Samples." *Diseases of Aquatic Organisms* 59: 85–91.

Schmeer, M. R., D. Horton, and A. Tanimura. 1966. "Mercenene, a Tumor Inhibitor from *Mercenaria*: Purification and Characterization Studies." *Life Science* 5: 1169–77.

Scozzari, Lois. 1995. "The Significance of Wampum to Seventeenth-Century Indians in New England." *Connecticut Review:* 59–69.

Shumway, Sandra E., and G. Jay Parsons. 2006. *Scallops: Biology, Ecology and Aquaculture.* Developments in Aquaculture and Fisheries Science, no. 35. Amsterdam: Elsevier.

Smolowitz, R., D. Leavitt, B. Lancaster, E. Marks, R. Hanselmann, and C. Brothers.

2001. Laboratory-Based Transmission Studies of Quahog Parasite Unknown (QPX) in *Mercenaria mercenaria*." *Journal of Shellfish Research* 20: 555–56.

Smolowitz, Roxanna, Dale Leavitt, and Frank Perkins. 1998. "Observations of a Protistan Disease Similar to QPX in *Mercenaria mercenaria* (Hard Clams) from the Coast of Massachusetts." *Journal of Invertebrate Pathology* 71: 9–25.

Stokes, N. A., L. M. Ragone Calvo, K. S. Reece, and E. M. Burreson. 2002. "Molecular Diagnostics, Field Validation, and Phylogenetic Analysis of Quahog Parasite Unknown (QPX), a Pathogen of the Hard Clam *Mercenaria mercenaria*." *Diseases of Aquatic Organisms* 52: 233–47.

Stokes, Nancy A., Mark E. Siddall, and Eugene M. Burreson. 1995. "Detection of *Haplosporidium nelsoni* (*Haplosporidia: Haplosporidiidae*) in Oysters by PCR Amplification." *Diseases of Aquatic Organisms* 23: 145–52.

Tamburri, M. N., R. K. Zimmer, and C. A. Zimmer. 2007. "Mechanisms Reconciling Gregrious Larval Settlement with Adult Cannibalism." *Ecological Monographs* 77, no. 2: 255–68.

Tehanetorens. 1999. *Wampum Belts of the Iroquois*. Summertown, Tenn.: Book Publishing Company.

Thoreau, Henry David. [1865] 2004. *Cape Cod*. Reprinted in *The Writings of Henry D. Thoreau*. With an Introduction by Robert Pinsky. Princeton: Princeton University Press.

Timmons, Michael, G. Gregg, D. Baker, J. Regenstein, M. Schreibman, P. Warner, D. Barnes, and K. Rivara. 2004. *New York Aquaculture Industry: Status, Constraints and Opportunities. A White Paper*. Cornell Cooperative Extension and New York Sea Grant, Cornell University Department of Biological and Environmental Engineering. www.bee.cornell.edu/cals/bee/outreach/aquaculture/upload/2New_York_Aquaculture_Industry_White_Paper.pdf.

True, Gabriella. 2003. "Culinary Discovery: The Tastes of Rhode Island." On Senior Women Web, http://www.seniorwomen.com/hs/articles/true/articlesTrueRhode.html.

Union Oyster House. n.d. "History of the Union Oyster House." Union Oyster House website, www.unionoysterhouse.com/Pages/history.html.

U.S. Army Corps of Engineers. 2007. "Draft Decision Document Nationwide Permit D." USACE website, www.usace.army.mil/cw/cecwo/veg/nwp/NWP_D_2007_draft.pdf.

USDA. 2005. *Census of Aquaculture.* United States Department of Agriculture. National Agricultural, Statistics Services.

Vanhaeren, Marian, Francesco d'Errico, Chris Stringer, Sarah L. James, Jonathan A. Todd, and Henk K. Mienis. 2006. "Middle Paleolithic Shell Beads in Israel and Algeria." *Science* 312: 1785–88.

Whitaker, D. *PSP Monitoring.* 2007. Massachusetts Department of Fish and Game, Division of Marine Fisheries, www.mass.gov/dfwele/dmf/programsandprojects/pspmoni.htm#shelsani.

White, Marie E., and Elizabeth A. Wilson. 1996. "Predators, Pests and Competitors." In *The Eastern Oyster* "Crassostrea virginica," edited by Victor S. Kennedy, Roger I. E. Newell, and Albert F. Eble, 559–75. College Park, Md.: Maryland Sea Grant College.

Wood, R. D. 1962. "Codium Is Carried to Cape Cod." *Bulletin of the Torrey Botanical Club* 89, no. 3: 178–80.

INDEX

Page numbers in **bold** indicate illustrations or tables